SOURCE CODE OPTIMIZATION TECHNIQUES FOR DATA FLOW DOMINATED EMBEDDED SOFTWARE

Source Code Optimization Techniques for Data Flow Dominated Embedded Software

by

Heiko Falk
University of Dortmund,
Germany

and

Peter Marwedel
University of Dortmund,
Germany

KLUWER ACADEMIC PUBLISHERS
BOSTON / DORDRECHT / LONDON

A C.I.P. Catalogue record for this book is available from the Library of Congress.

ISBN 1-4020-2822-9 (HB)
ISBN 1-4020-2829-6 (e-book)

Published by Kluwer Academic Publishers,
P.O. Box 17, 3300 AA Dordrecht, The Netherlands.

Sold and distributed in North, Central and South America
by Kluwer Academic Publishers,
101 Philip Drive, Norwell, MA 02061, U.S.A.

In all other countries, sold and distributed
by Kluwer Academic Publishers,
P.O. Box 322, 3300 AH Dordrecht, The Netherlands.

Cover design by Alexandra Nolte

Printed on acid-free paper

Printed in the Netherlands.

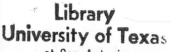

Contents

List of Figures

List of Tables

To my unborn child.

Acknowledgments

This book represents a revised version of my doctoral thesis submitted to the Department of Computer Science at the University of Dortmund in March 2004. My gratitude goes to my advisor Prof. Dr. Peter Marwedel for giving me the opportunity and the freedom to perform research in the area of high-level source code transformations. Furthermore, I would like to thank my co-referee Prof. Dr. Peter Padawitz for his efforts.

Many thanks go to my colleagues at the Embedded Systems Group of the Department of Computer Science at the University of Dortmund. Especially, I want to thank Dr. Markus Lorenz, Manish Verma and Lars Wehmeyer for all the valuable discussions and their effort in proofreading this book.

The lovely cover design of this book was created by Alexandra Nolte.

Contributions to the results described in this book were made by several people, whose help is gratefully acknowledged. In particular, Francky Catthoor, Cédric Ghez, Miguel Miranda and Martin Palkovic from the IMEC research center at Leuven (Belgium) should be mentioned here.

My family reminded me day by day that important parts of life take place outside the office. They gave me the love, patience and understanding to obtain the Ph.D. degree and finally to write this book.

Dortmund, August 2004   Heiko Falk

Foreword

This book focuses on source-to-source code transformations that remove addressing-related overhead present in most multimedia or signal processing application programs. This approach is complementary to existing compiler technology. What is particularly attractive about the transformation flow presented here is that its behavior is nearly independent of the target processor platform and the underlying compiler. Hence, the different source code transformations developed here lead to impressive performance improvements on most existing processor architecture styles, ranging from RISCs like ARM7 or MIPS over Superscalars like Intel-Pentium, PowerPC, DEC-Alpha, Sun and HP, to VLIW DSPs like TI C6x and Philips TriMedia. The source code did not have to be modified between processors to obtain these results. Apart from the performance improvements, the estimated energy is also significantly reduced for a given application run.

These results were not obtained for academic codes but for realistic and representative applications, all selected from the multimedia domain. That shows the industrial relevance and importance of this research. At the same time, the scientific novelty and quality of the contributions have lead to several excellent papers that have been published in internationally renowned conferences like e. g. DATE.

This book is hence of interest for academic researchers, both because of the overall description of the methodology and related work context and for the detailed descriptions of the compilation techniques and algorithms. But the realistic applications should also make it interesting for senior design engineers and compiler / system CAD managers in industry, who wish to either anticipate the evolution of commercially available design tools over the next few years, or to make use of the concepts in their own research and development.

In particular, several contributions have been collected in this book. Based on a good study of the related work in the focused area, it is concluded that especially for expressions found e. g. in address arithmetic, the control flow surrounding these expressions is extremely important for the final performance on the target architecture. If this control flow is not sufficiently optimized, the overall system performance can be affected in a very negative way. In order to avoid and remedy this, it is crucial to apply source code transformations which remove this overhead. The author has proposed and worked out novel ways of applying loop nest splitting and advanced code hoisting in order to achieve this objective. He has developed techniques to steer these source code transformations within the large search space that is present in realistic applications. He has also substantiated the validity and large impact of these techniques on several realistic applications for a multitude of modern processor architecture styles. In addition, he also looked at additional transformations like ring buffer replacement that are more specific but also show promising gains in particular situations.

The material in this book is partly based on work in a Ph.D. research co-operation between the University of Dortmund and the DESICS division at IMEC. We believe this has contributed to the results by combining the expertise and strengths from both types of research environments. It has been a pleasure for me to co-operate with the author and my colleagues at Dortmund in this area and I am confident that the material in this book can be of value to you as a reader.

Leuven, August 2004   Prof. Francky Catthoor
                                                       IMEC Fellow
                                                       Leuven, Belgium

Chapter 1

INTRODUCTION

In general, two different classes of digital systems exist: general-purpose systems and special-purpose systems. *General-purpose systems* are, for example, traditional computers ranging from PCs, laptops, workstations to supercomputers. All these computers can be programmed by a user and support a variety of different applications determined by the executed software. General-purpose systems are not developed for special, but for general applications. During the eighties and early nineties, these general-purpose systems were the main drivers for the evolution of design methods for computer and compiler architectures. This situation is changing, because several new applications (e. g. UMTS) have come up which can not be implemented efficiently by general-purpose systems.

For this reason, *special-purpose systems* developed to fulfill one special, fixed task gain more and more importance. Most of these systems are configured once and work independently afterwards. The user has limited access to program these systems. These special-purpose systems can be characterized by the following features [Marwedel, 2003]: They...

- are usually embedded in other products. Hence, the notion of *embedded systems* is commonly used.

- work as stand-alone devices.

- are very often I/O intensive.

- are infrequently reprogrammed. Their functionality is mostly fixed. In research, the dynamic reprogramming of embedded systems at runtime is becoming more and more important, but this feature can not yet be found in today's systems.

1

Figure 1.1.   Examples of typical Embedded Systems

- are extremely cost-, power- and performance sensitive.

- often have hard real-time, reliability and correctness constraints.

- very often are reactive, responding frequently to external inputs.

Some typical examples of embedded systems are depicted in figure 1.1. The importance of embedded systems is highlighted by the fact that in 1995, more than the half of sold 32-bit processors (57%) were used for embedded systems, and only the remaining 43% for computing applications [Paulin et al., 1997]. Furthermore, 8- or 16-bit processors dominate the embedded system market with 95% in volume, in contrast to general-purpose systems.

EXAMPLE 1.1

The current model of the BMW 745i sedan includes around 70 different processors [Givargis, 2004]:

- *53 8-bit microprocessors,*

- *7 16-bit microprocessors and*

- *11 32-bit microprocessors.*

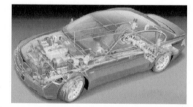

The total amount of more than 60 megabytes of software is executed in this car [Saad, 2003], corresponding to 2,000,000 lines of assembly code.

Nowadays, the contribution of automotive electronic components to the total added value of a vehicle amounts to to 35%. It is expected that 90% of all innovations in vehicles will be based on electronic systems in the future [Weber, 2003].

The design process of embedded systems is always driven by the goal to maximize the system value while minimizing the cost. In order to combine a growing system complexity with an increasingly short time-to-market, new system design technologies have emerged based on the paradigm of *embedded programmable processors*. This class of processors is optimized for certain application areas (such as audio and video applications) or even for certain applications. However, optimized processor architectures used in embedded systems are only able to meet all cost-, power- and performance-criteria when executing highly optimized and efficient machine code.

1.1 Why Source Code Optimization?

Due to the need for very efficient machine code and the lack of compilers generating efficient code, assembly-level software development for embedded processors has been very common in the past. However, both embedded applications and embedded processors are getting more and more complex, and the development of highly optimizing compilers allowed the replacement of assembly level programming. Nowadays, it can be observed that the focus of research moves from compiler technologies towards optimizations applied prior to compilation at the source code level. The reasons for this shift towards precompilers are manifold.

First, the application of optimizations at the source code level makes these techniques inherently portable. Provided that a standardized programming language is used as in- and output for transformation, an optimized source code can be fed into any available compiler for the programming language, irrespective of actual processor architectures. In this book, the programming language ANSI-C [Kernighan and Ritchie, 1988] is used for source code optimization due to its wide dispersion in the domain of embedded systems.

Second, the correctness of an optimization can be checked more easily at the source code level. For this purpose, an unoptimized source code and its optimized version only need to be compiled and executed. If both versions behave differently, the optimization step was erroneous. This approach is very efficient because compilation and execution of both code versions can be performed using any available computer and compiler due to the portability of the source codes. In contrast, a verification at the compilation level requires the execution of the programs on the dedicated target processor. In the context of embedded systems, the execution speed of embedded processors is much lower than that of desktop workstations resulting in larger runtimes of the compiled programs. Additionally, embedded processors are often unavailable in hardware when developing optimizations, so that time consuming simulations are necessary.

Third, research on optimization techniques is easier to perform at the source code level than at machine code level. When investigating the effects of an optimization technique at a low level of abstraction, this can in most cases only be done if the technique to be studied is already fully implemented. In contrast,

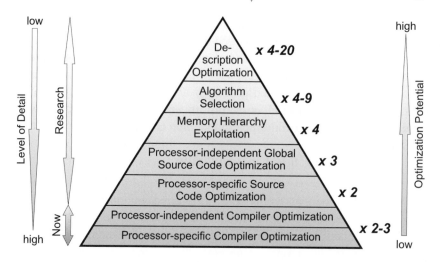

Figure 1.2.   Abstraction Levels and achievable Improvements of Code Optimizations

source codes are human-readable and -understandable due to their higher level of abstraction. This makes it possible to apply an optimization manually in order to explore its effects. This way, it can be evaluated very quickly if it is worthwhile to study an optimization or not. Only if the manual experiments lead to satisfying results, the efforts of a full automation are needed.

However, the most important reason for research on source code optimizations is that they are highly beneficial because of their high level of abstraction. Due to its importance, this correlation is explained in detail in the following section 1.1.1. In order to further demonstrate the need for research on source code transformations, the design flow employed by a traditional code generation and optimization approach is described in section 1.1.2 along with its deficiencies. Finally, in section 1.1.3, different scopes of code optimization are presented. It is shown at which level commonly known optimizations which are integrated into today's optimizing compilers are applied.

1.1.1   Abstraction Levels of Code Optimization

Research has shown that massive improvements are achievable when considering a very high level of abstraction for the application of optimization techniques. The correlation between the abstraction level and corresponding potential gains is depicted in figure 1.2.

At the bottom of the pyramid representing the code optimization process, all kinds of optimizations integrated in today's state of the art compilers can be found. These optimizations can be divided into processor-specific and processor-independent techniques as explained in more detail in the follow-

ing section. Typically, gains of a factor of two or three can be achieved at
this level [Muchnick, 1997, Leupers, 2000][1]. In the area of processor-specific
source code optimizations (e. g. exploitation of compiler intrinsics), improvements of a factor of two have been reported in [Vander Aa et al., 2002, Wagner
and Leupers, 2001, Pokam et al., 2001, Coors et al., 1999]. The experimental
results provided by this book show that improvements of a factor of three can
be achieved by the application of processor-independent global optimizations
at the source code level. The same order of magnitude has also been observed
in [Ghez et al., 2000]. Even higher gains can be achieved by an optimized exploration of memory hierarchies. The minimization of data transfers between
functional units and memories as described in [Hu, 2001] followed by address
code optimizations leads to gains of a factor of four. At an even higher level of
abstraction, different functionally equivalent implementations of an algorithm
of an application are studied. In [Püschel et al., 2001], a framework for the selection of the most efficient realization of an algorithm is presented leading to
approximate gains of a factor of four. At the top level of the pyramid, optimizations of the description of a program can be found. These techniques include
the modification of used data types (e. g. replacement of double precision float-
ing point numbers by single precision floats or integers) or the replacement of
complex formulas by simpler approximations. At this level, it is still allowed
to change the behavior of a program (within a certain tolerance) by a transformation, whereas the functional equivalence has to be maintained at all other
levels. For this reason, such approaches have proven to be highly effective;
improvements by a factor of four up to twenty have been published in [Sinha
et al., 2000, Peymandoust et al., 2002, Hüls, 2002].

   As can be seen from this overview, optimizing compilers are effective in
generating efficient machine code for a given processor architecture. Since
all optimizations beyond the scope of today's state of the art compilers are
an area of ongoing research, an integration of such techniques into commonly
known compilers – although generally possible – is currently inexistent. For
this reason, optimizations at a high abstraction level are expressed as source
code transformations.

1.1.2   Survey of the traditional Code Optimization Process

   A compiler designed to produce efficient object code includes optimizers. In
figure 1.3, the source code is translated to a medium-level intermediate code[2],
and optimizations that are largely processor-independent are performed on

---

[1] All references cited in this paragraph are presented thoroughly in chapter 2.
[2] See chapter 4 for more information on intermediate representations (IR).

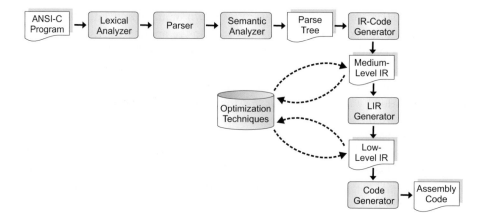

Figure 1.3.   Structure of Optimizing Compilers

it [Leupers and Marwedel, 2001]. Examples for such optimizations are [Muchnick, 1997]:

- **Common subexpression elimination**
  A computation is redundant at some point in a program, if that computation has already been performed at an earlier program point. Such redundancies can be eliminated by keeping the result of the first occurrence of the computation in a variable and reusing the variable instead of re-computing the value.

- **Dead code elimination**
  Any computation that generates a result which is guaranteed to be never used in the program (so called "dead code") can be eliminated without affecting the program semantics.

- **Loop-invariant code motion**
  Computations contained in a loop whose results do not depend on other computations performed in that loop are said to be loop-invariant. Consequently, these computations can be moved outside of the loop code in order to increase performance.

- **Constant folding**
  Computations that are guaranteed to result in constants can be moved from program runtime to compile-time. In this case, the compiler "folds" computations on constants to single constants already when generating assembly code.

Hereafter, the improved processor-independent code is translated to a low-level form and further optimizations that are mostly processor-specific are applied to it. These optimizations mainly deal with the efficient mapping of a program to machine-specific assembly instructions. This process which is often called *code generation* is usually decomposed into several phases for which efficient optimization techniques have been proposed [Leupers, 2000]:

- **Code selection**
  In this phase, it is decided which assembly instructions will be used to implement code represented by a low-level intermediate representation. The optimization goal is to select a set of assembly instruction instances whose accumulated costs, with respect to a given metric, are minimal.

- **Data routing**
  After operations are bound to functional units of the target processor during code selection, data routing [Rimey and Hilfinger, 1988, Lanneer et al., 1994, Leupers, 1997] deals with transporting data between functional units so as to minimize required transport operations.

- **Register allocation**
  The register allocation phase determines where the results of computations are stored, either in memory or in a processor-internal register. Since the number of available physical registers is limited, register allocation decides which program values need to share a register such that the number of registers simultaneously required at any program point does not exceed the physical limit. The optimization goal is to hold as many values as possible in registers in order to avoid expensive memory accesses.

- **Instruction scheduling**
  This phase decides when instructions will be executed at program runtime. During this optimization phase, dependencies between instructions as well as limited processor resources and pipelining effects have to be taken into account. The optimization goal can be to minimize the required execution time.

Using this structure of optimizing compilers (see figure 1.3), the two lowest levels depicted in figure 1.2 are implemented. Although being very close to the processor architecture, compilers are often unable to map a source code efficiently to the actual platform. This is illustrated by means of the code depicted in figure 1.4.

This source code fragment taken from GSM and G.723.1 software shows an implementation of a *saturating addition*. As can be seen, the result of an addition is stored in a variable first. Afterwards, the saturation of this result – i. e. limiting the result to to given minimal and maximal constant values –

```
result = var1 + var2;

if (((var1 ^ var2) & 0x80000000) == 0)
    if ((result ^ var1) & 0x80000000)
        result = var1 < 0 ? 0x80000000 : 0x7fffffff;
```

Figure 1.4.   Implementation of a saturating Addition

is realized using several cascaded *if*-statements. Although typical embedded processors (e. g. TI C6x or Philips TriMedia) can execute a saturating addition using one machine instruction, compilers do not identify this behavior in the source code. Hence, very inefficient control flow dominated machine code is generated during compilation.

Currently, the highest level of optimizations performed by compilers is formed by loop optimizations. The goal of these optimizations (e. g. loop interchange, tiling or distribution [Bacon et al., 1994, Muchnick, 1997]) is to restructure the iteration traversal of loops so as to improve the locality of memory accesses. This way, the efficiency of data caches can be improved due to less cache misses. These optimizations only change the order in which memory accesses are performed and do not modify the total number of memory accesses. Techniques aiming at a higher level where data structures and algorithms are transformed in order to reduce the number of memory accesses are still beyond the scope of compilers.

1.1.3   Scopes for Code Optimization

In this section, it is shown which scopes are typically used within compilers for code optimization. The following material bases on state of the art literature on optimizing compilers [Aho et al., 1986, Bacon et al., 1994, Muchnick, 1997] and thus reflects the main principles of today's commonly used compilers. Of course, some compilers exist behaving differently than described here (e. g. non-commercial research compilers), but the major trend-lines are mirrored adequately.

Generally, optimizations can be applied to a program at different levels of granularity. As the the scope of a transformation is enlarged, the cost of analysis generally increases. Some useful gradations of complexity are [Bacon et al., 1994]:

- **Statement**
  Arithmetic expressions are the main source of potential optimization within a statement.

- **Straight-line code**
  This is the focus of early optimization techniques. The advantage for anal-

ysis is that there is only one entry point, so control transfers need not be considered in tracking the behavior of the code.

- **Innermost loop**
  To target high-performance architectures efficiently, compilers need to focus on loops. It is common to apply loop manipulations only in the context of the innermost loop.

- **Perfect loop nest**
  A loop nest is a set of loops one inside the next. The nest is called a perfect nest if the body of every loop other than the innermost consists of only the next loop in the nest. Because a perfect nest is more easily organized, several optimizations apply only to perfect nests.

- **General loop nest**
  Any loop nesting, perfect or not.

- **Procedure**
  Some optimizations, memory access transformations in particular, yield better improvements if they are applied to an entire procedure at once. The compiler must be able to manage the interactions of all the control transfers within the procedure. The standard and rather confusing term for procedure-level optimization in the literature is *global optimization*.

- **Inter-procedural**
  Considering several procedures together often exposes more opportunities for optimization. In particular, procedure call overhead is often significant and can sometimes be reduced or eliminated with inter-procedural analysis.

By far most of the optimizations performed by a compiler operate at the statement-level, e. g. constant propagation and folding, copy propagation, dead code elimination [Muchnick, 1997]. A survey of actual compiler infrastructures (see chapter 4) has shown that compilers frequently represent programs using building blocks of straight-line code. As a consequence, many powerful optimizations like common subexpression elimination or algebraic reassociation are applied only under consideration of these blocks so that optimization potential beyond straight-line code frequently is not explored.

Additionally, this model of blocks of straight-line code implies that costly analysis has to be performed in order to gather information about loops. If this is done, it is a common simplification to restrict the focus of loop optimizations to the innermost loops of a perfect loop nest. Optimizations that are able to deal with e. g. memory accesses in outer loops guarded by complex conditions and which are still able to improve the locality of a program are uncommon.

At the level of procedures, compilers typically perform transformations in order to reduce to procedure-call overhead, e. g. procedure inlining or tail recursion elimination. Generally, the procedural scope is not used for the transformation and replacement of entire algorithms.

Concluding, it can be summarized that today's commonly used compilers which are designed for actual processors are very limited with respect to their scopes and abstraction levels of optimization (cf. also chapter 4 on intermediate representations). The models of program representation used by compilers are restricted in such a way that important information about the overall structure of a program is not considered at all or is not used for an in-depth exploration of global optimization potential. These circumstances underline the need of new optimization techniques focusing on the higher levels of a program's structure. Efforts are made in order to integrate such techniques into research compilers (e. g. SUIF [Wilson et al., 1995], see section 4.3.1), but the research and development of these optimizations at the source code level appears to be promising in order to bypass the limitations imposed by compilers.

1.2   Target Application Domain

The source code optimization techniques presented here focus on a specific target application domain. As mentioned in the title of this book, the main target domain consists of embedded signal and data processing systems which deal with large amounts of data (*"data flow dominated"*). This target domain can be subdivided into three classes: multimedia applications (e. g. video processing, medical imaging), front-end telecommunication applications (e. g. GSM, G.723.1) and network component applications (e. g. ATM or IP network protocols). The main focus of this book lies on multimedia applications. The main characteristics of this class of applications are the following [Catthoor et al., 2002]:

- **Loop nests**
  Multimedia applications typically contain many deep loop nests for processing generally multidimensional data. Due to the rectangular shape of images and video streams, the loop bounds are often constant, or at least dependent on other iterators. This way, it can be assumed that loop nests are fully statically analyzable. The loops are typically quite irregularly nested though.

- **Mostly statically analyzable affine conditions and indices**
  Most of the conditions and indexing are affine functions of the loop iterators.

- **Large multidimensional arrays**
  Multimedia applications mainly work on multidimensional arrays of data. Large one-dimensional arrays are also often encountered.

- **Statically allocated data**
  Most data is statically allocated. Its exact storage location can be optimized at compile time.

In addition to the above, the data flow dominated applications considered in this book are typically characterized by high throughput and low power requirements. Furthermore, timing constraints have to be taken into account. Finally, the code of this kind of embedded software is often complex and long, consisting of data-dominated modules with large impacts on cost issues.

Telecommunication applications generally have the same characteristics as multimedia applications, with the exception that the loop bounds may be partly varying. Network component applications, on the other hand, contain almost no deep loops and have mainly data dependent conditions and indices. Moreover, most of the data is organized in dynamic tables and lists allocated at run-time.

Especially in the domain of multimedia applications, the trend towards re-programmable computing adding a more dynamic functionality to embedded systems can be observed recently. Based on embedded programmable processors, embedded applications are becoming more and more complex and often consist of several tasks that are created and terminated dynamically. Additionally, the dynamic download of application code during the runtime of an embedded system is studied nowadays [Kuacharoen et al., 2003]. This no longer fixed functionality of embedded data flow dominated systems does not prevent the optimizations presented in this book to be used. This is due to the fact that they are applied once to every individual task of an application prior to compilation. The execution of the highly optimized machine code of a task can then be steered dynamically during runtime.

1.3 Goals and Contributions

The goals of the work presented in the following chapters are:

- Development of novel optimization techniques which

- can be automated using the material presented in this book,

- are applied at the source code level to embedded applications, and

- which aim at minimizing execution times and energy dissipation.

As mentioned previously, the techniques presented in this book can be classified as processor-independent global source code optimizations (cf. figure 1.2). Although motivated by different architectural properties (e. g. instruction pipeline or address generation units), these optimization techniques do not take e. g. different pipeline layouts or the potential support of modulo addressing into account. For this reason, one of the main contributions of these optimizations is their independence of actual processor architectures. The provided experimental results demonstrate that various different processor architectures benefit by comparable orders of magnitude. These analogies between different processors underline the high quality of the optimizations independently of actual architectures.

The optimizations presented in the following bridge the gap between the techniques effectively exploring memory hierarchies and those focusing on a dedicated hardware. A survey of related work (see chapter 2) has shown that especially in this area, many issues have not yet been addressed. For example, the intensive optimization of data transfers and memory hierarchies implicitly leads to a degradation of the control flow of an embedded application. For this reason, another important contribution of this work is its concentration on control flow aspects in data flow dominated embedded software.

In order to meet the goals of automated optimization of runtimes and energy consumption, basic optimization strategies which are rather unconventional in classical compiler construction are utilized. These include genetic algorithms and geometrical polytope models. Using such powerful techniques allows to solve complex code optimization problems for which good heuristics are out of sight. Although these basic strategies are quite complex, all source code optimizations have proven to be highly efficient so that times elapsed during code optimization are negligible.

As will be shown experimentally in each chapter, these approaches for source code optimization achieve significant improvements in code quality. Code quality is quantified using numerous existing processors which are available in real hardware. The collection of results bases on measurements of the physical processor hardware. Time consuming and imprecise software simulations were avoided to a large extent. Since another contribution of this book is to provide optimization techniques that work in practice, mostly all optimizations are fully automated and applied to real-life embedded software and processors.

1.4   Outline of the Book

The remainder of this book is organized as follows:

- **Chapter 2** gives a general overview of the related work in the domain covered by this book.

- **Chapter 3** presents the basic optimization strategies utilized by source code optimization techniques. Additionally, the benchmarking methodology is explained.

- **Chapter 4** classifies intermediate formats used for the the representation of programs by compilers. Different existing models are presented, and an experimental comparison of two intermediate representations is given so as to select the one which is most suitable for source code optimization.

- **Chapter 5** contains a detailed description of the first code optimization technique called "loop nest splitting". Its goal is to generate a homogeneous control flow in deeply nested loops of data flow dominated applications.

- **Chapter 6** treats a new optimization technique called "advanced code hoisting". Here, the elimination of complex index expressions used for memory addressing under consideration of control flow aspects is discussed.

- **Chapter 7** describes the replacement of small arrays which are accessed in a circular manner by scalar variables. Combined with loop nest modifications, this technique forms a simultaneous optimization of memory accesses and addressing.

- **Chapter 8** concludes the book and points out the most interesting areas for future research.

Chapter 2

EXISTING CODE OPTIMIZATION TECHNIQUES

Several research groups in the area of software synthesis for (embedded) programmable processors have developed techniques supporting different aspects of important optimization problems. In this chapter, an overview and a classification of existing work is given.

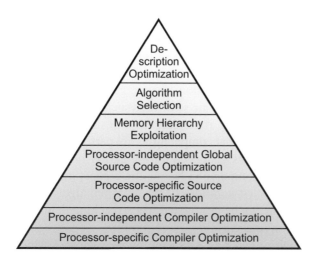

Figure 2.1. Abstraction Levels of Code Optimizations

Figure 2.1 illustrates the different abstraction levels to which code optimizations are commonly applied. According to this diagram, related contributions are presented, starting with the optimization of descriptions described in section 2.1 and ending with standard compiler optimizations in section 2.6.

This chapter does not claim to be an exhaustive record of related work on code optimization techniques. Instead, only an impression of the complex problems arising during software synthesis and of effective solutions is provided. Whenever necessary, more detailed references to particular subjects are included in those chapters of this book describing individual optimization techniques.

2.1   Description Optimization

As mentioned previously, only optimizations performed at the description level are allowed to modify the functional behavior of an application within certain tolerances. For all other levels, an optimized application code has to behave exactly like the unoptimized application. At the description level, the *SymSoft* tool flow consisting of two steps is presented in [Peymandoust et al., 2002]. First, it is checked whether the software data representation matches the facilities of the processor hardware. Most embedded processors support only fixed-point computation, but many multimedia algorithms utilize floating-point operations. If such a mismatch between software and hardware is detected, a conversion of data formats takes place. In the second step, polynomial approximation techniques are applied to the most energy consuming parts of a source code. For instance, *symbolic algebra* transforms the expression $y = \cos(x) + \cos(2 * x) + \cos(3 * x)$ to $y = 3 + (-7 + \frac{49}{12}x^2)x^2$. This transformation process is executed in an interactive loop so that the designer can check whether the accuracy of a polynomial formula is sufficient or not. Using this approach, reductions of runtimes of up to 62% and of energy dissipation of 77% are reported.

In [Sinha et al., 2000], the notion of *energy scalable computation* is introduced. In this work, algorithms are structured such that computational accuracy can be traded off with energy requirements. For this purpose, algorithms are transformed in such a way that computations are performed incrementally, basing on intermediate results generated during earlier iterations. Care is taken that those parts of the computation having the highest impact on accuracy are performed during the first iterations of the rewritten algorithm. This way, the iterative computation can be terminated when reaching defined thresholds for energy consumption or accuracy.

The *replacement of complex data types* by simpler ones which are directly supported by the processor hardware is studied in [Hüls, 2002] by means of a complete MPEG-2 software. The replacement of double precision floating-point numbers by single precision floating-point numbers has proven not to affect the accuracy of the application. Large savings of energy consumption (66%) and runtimes (65%) have been observed for an ARM7 embedded RISC processor. Due to the fact that the considered processor does not support floating-point numbers in hardware, these variables have been replaced by integer numbers in a second step. Huge gains of around 95% for both runtimes and energy consumption are reported. It is mentioned that the quality loss is noticeable, but trade-offs are not discussed.

2.2 Algorithm Selection

The *SPRIAL* framework for automatic generation and selection of DSP algorithms is presented in [Püschel et al., 2001]. SPIRAL is a system generating libraries for digital signal processing algorithms which are optimized for a given processor architecture. In this framework, the user can specify a computation to be performed. This specification is called "transform". A formula generator is able to iteratively create several equivalent formulas for a given transform. All these different formulas compute the transform, but they differ in the data flow required during computation, causing a wide range of actual runtimes. A generated formula is exported as C or Fortran program and fed into the search module of the SPIRAL framework. Here, the runtime performance of a generated source code algorithm is evaluated and fed back into the formula generator. This performance measure is used by the formula generator so as to create an equivalent formula with higher performance during the next iteration.

Due to the very close relationship between an algorithm and the data structures it is manipulating, any optimization of the layout of an algorithm's data structures can be interpreted as algorithm transformation as well. In [Daylight et al., 2002], an approach for the reorganization of data structures is presented. Starting from a set of three different primitive data structures, combination and key splitting steps are applied. These techniques lead to the generation of a multi-layered hierarchy of primitive data structures. This hierarchy is built in such a way that data retrieval and manipulation is performed easily consuming least energy. Hereafter, this hierarchy is partitioned in order to place fragments of the unpartitioned data structure onto different memories of a given memory hierarchy. Using this approach, the energy consumption of the generated hierarchies of data structures can be traded off with the corresponding memory space consumption. For a 3D computer game, a maximum energy consumption reduction by a factor of 8.7 was observed.

2.3 Memory Hierarchy Exploitation

The optimization of the global memory accesses of an application is crucial for achieving low power systems, because it has been shown that between 50% and 80% of the power in embedded multimedia systems is consumed by memory accesses. In [Catthoor et al., 2002], the so-called *Data Transfer and Storage Exploration* methodology (DTSE) is proposed. The goal of DTSE is on the one hand to minimize the storage requirements of embedded applications so as to reduce the absolute amount of memory needed to execute the application. On the other hand, the locality of data memory accesses is optimized at a very

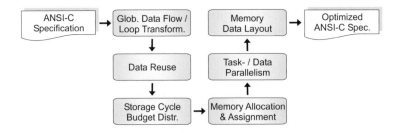

Figure 2.2. Overview of the DTSE Methodology

high global level in order to achieve a high utilization of small energy-efficient memories which are close to the processor.

The DTSE methodology consists of several steps (cf. figure 2.2). In the beginning, a memory oriented data flow analysis is performed followed by global data flow and loop transformations in order to reduce the required amount of background memories. Hereafter, data reuse transformations exploit a distributed memory hierarchy. Storage cycle budget distribution is performed in order to determine the bandwidth requirements and the balancing of the available cycle budget over the different memory accesses in the algorithmic specification. The goal of the memory hierarchy layer assignment phase is to produce a netlist of memory units from a memory library as well as an assignment of signals to the available memory units [Brockmeyer et al., 2003]. For multi-processor systems, a task- / data-parallelism exploitation step [Hu, 2001] takes place minimizing the communication and storage overhead induced by the parallel execution of subsystems of an application. Finally, data layout transformations (e. g. in-place mapping) help to increase the cache hit rate and to reduce the memory size by storing signals with non-overlapping lifetimes in the same physical memory location.

DTSE has proven to be highly effective in minimizing the energy cost related to data memory hierarchies. For several typical embedded applications, reductions of main memory accesses, cache accesses and energy consumption between 50% and 80% have been reported. Recently, the research on memory hierarchy transformations has led to the foundation of first spin-offs selling commercial products for source code optimization [PowerEscape Inc., 2004].

There exist several other works related to the efficient utilization of memories [Kim et al., 2000, Fraboulet et al., 1999, Kandemir, 2002]. They all differ from the DTSE methodology in the point that the global data flow of an application remains unchanged. For a given sequence of accesses to arrays of a fixed size, these techniques aim at rearranging the access sequence in an optimized way. In contrast, the global view of DTSE allows explicitly to modify sizes and accesses of arrays. The references given above are presented in more detail in section 5.2 in the context of the loop nest splitting optimization.

In [Panda et al., 1999], an efficient use of a memory hierarchy was presented by placing the most commonly used variables onto a small energy-efficient *scratchpad* memory. A further approach published in [Sjödin et al., 1998] places variables onto a scratchpad by static analysis and shows that this is sufficiently precise and no dynamic analysis is needed. The dynamic copying of array parts was studied in [Kandemir et al., 2001]. However, the preconditions imposed by this algorithm are very restrictive so that only a limited class of applications can be analyzed. An approach for statically placing both data and program parts onto a scratchpad memory basing on the analysis of instruction execution and variable accesses is presented in [Steinke et al., 2002b]. In that paper, only whole variables are considered at a time. Consequently, a large non-scalar variable (i. e. an array) could either be placed onto the scratchpad as a whole or not, potentially leading to a sub-optimal memory utilization.

2.4 Processor-independent Source Code Optimizations

The DTSE steps described above will typically significantly increase the addressing complexity of transformed applications, e. g. by introducing more complex loop bounds, index expressions, conditions etc. Since this computational overhead is neglected by the DTSE methodology, the memory optimized system, despite being more power efficient, will suffer a severe performance degradation. Typical examples show that the runtime performance degrades by a factor of five or six after DTSE [Danckaert et al., 1999, Ghez et al., 2000].

Parts of the additional complexity introduced by DTSE can be removed by source code optimizations for *address optimization* (ADOPT) [Ghez et al., 2000]. In this context, an algebraic cost minimization is performed first minimizing the number of operation instances. The goal of this step is to find factorizations of addressing expressions in order to be able to reuse computations as much as possible. The factorization step is performed using a so-called "mountain-trip method" [Janssen, 2000], whereas the reuse of expressions bases on conventional common subexpression elimination techniques.

A very high penalty is caused by the DTSE transformations by the frequent insertion of non-linear addressing expressions containing modulo or division operators. Since these operators are usually not supported by the instruction sets of embedded processors, they have to be realized by calls to software routines of system libraries. For this reason, the optimization of non-linear addressing is another major goal of the ADOPT transformations. Modulo and division can be replaced by variables serving as accumulators in the following way:

```
                                        int jdiv6=0, jmod6=0;

  for (j=0; j<n; j++)              for (j=0; j<n; j++, jmod6++) {
    a[j/6][j%6] = ...;       →       if (jmod6 >= 6) {
                                       jmod6 -= 6; jdiv6++; }
                                     a[jdiv6][jmod6] = ...; }
```

After this kind of transformation, addressing computations only consist of simple increment and decrement operators which are supported efficiently by every processor. As a consequence, the ADOPT transformations are able to reduce runtimes by a factor of three.

Nevertheless, the increased amount of addressing arithmetic is not the only kind of overhead introduced by DTSE. Control flow aspects have not yet been targeted at all in the context of data optimization. For this reason, the source code optimization techniques presented in this book explicitly focus on this area of research.

2.5   Processor-specific Source Code Optimizations

In the area of processor-specific source code optimization, research focuses on the exploitation of *intrinsics* (also known as *compiler known functions*) to a large extent. Using intrinsics, particular features of a processor can be made visible to the programmer. The compiler maps a call to an intrinsic not to a regular function call, but to a fixed sequence of machine instructions. Thus, intrinsics can be considered as source-level macros without any calling overhead. For example, the complex code shown on the left hand side of figure 2.3 implementing a saturating addition can be replaced by an intrinsic using the TI C6x compiler. This way, the compiler is able to replace the entire C fragment by a single instance of the SADD machine instruction.

```
   result = a + b;

if (!((a ^ b) & 0x80000000)
   if ((result ^ a) & 0x80000000)    →   result = _sadd(a, b);
     result = a < 0 ? 0x80000000 :
                       0x7fffffff;
```

Figure 2.3.   Replacement of complex Code by a TI C6x Intrinsic

This approach is included in several state of the art optimizing compilers. In [Wagner and Leupers, 2001], a compiler for a network processor supporting complicated bit-packet addressing facilities is presented. To make use of these particular machine instructions, the compiler allows to replace source code fragments dealing with bit-masking and -shifting by intrinsics. But also for already existing compilers, the effects of using compiler known functions for typical signal processors are studied in literature [Vander Aa et al., 2002, Coors et al., 1999]. In all these references, improvements between 30% and 50% have been measured.

Intrinsics are also within the scope of the *SWARP* source code transformation framework [Pokam et al., 2001]. This work concentrates on the exploitation of multimedia instruction set extensions of the Philips TriMedia processor. In a first step, loop transformations are applied so as to rearrange the computations in a loop nest appropriately. Second, particular code fragments in the transformed loop nest are identified using pattern matching and are replaced by intrinsics of the TriMedia compiler.

A more general framework for source code transformation based on pattern matching is presented in [Boekhold et al., 1999]. The *C transformation tool* (CTT) allows the user to define a library of transformations. Every transformation consists of a description of a source code fragment, of rules specifying the legality of transformation application, and of a description of the transformed source code. This transformation tool is used to apply several standard loop optimizations (see the following section 2.6.1) at source code level in order to parallelize programs.

For global source code transformations supporting general code structures, where e. g. arbitrary numbers of loops and *if*-statements can occur, pattern matching has proven to be inadequate [Jakubowski, 2002]. Due to the fact that individual patterns have to be defined for e. g. loop nests of different depths, such approaches are too inflexible for being considered in this book.

2.6 Compiler Optimizations

Besides the standard processor-independent compiler optimizations already presented in section 1.1.2 (see also [Aho et al., 1986, Muchnick, 1997]), more specialized optimizations focusing on particular processor architectures have been proposed. In the following, emphasis is placed on loop optimizations originally stemming from the high-performance computing community (see section 2.6.1) and on code generation techniques for typical embedded processors (section 2.6.2).

2.6.1 Loop Optimizations for High-Performance Computing

In earlier times, the construction of optimizing compilers was pushed by so called *high-performance architectures* (e. g. mainframes and super-computers). Already in this period, the main focus of compilers was on loop optimizations [Bacon et al., 1994]. A sequence of transformations is performed to reorder loop iterations so as to maximize parallelism and memory locality.

Although several years have passed since this time, and despite the fact that embedded processors can hardly be compared with mainframe CPUs at first sight, optimizing compilers for embedded processors base on the same ideas. This is motivated by the fact that the exploitation of parallelism and locality is of utmost importance, because typical embedded processors offer *instruction-level parallelism* and complex memory hierarchies. Typical examples of such loop optimizations are taken from [Bacon et al., 1994].

Loop interchange exchanges the position of two loops in a perfect loop nest, generally moving one of the outer loops to the innermost position:

```
for (i=0; i<n; i++)                      for (j=0; j<m; j++)
   for (j=0; j<m; j++)          →           for (i=0; i<n; i++)
      a[i][j] = a[i-1][j]+b[i][j];             a[i][j] = a[i-1][j]+b[i][j];
```

Loop interchange may be performed to enable vectorization by interchanging an outer loop whose iterations depend on each other with an inner one which is independent. If a loop nest allows parallelization already before loop interchange, this optimization can be used in order to improve vectorization by moving the independent loop with the largest range to the innermost position.

Loop unrolling replicates the body of a loop some number of times called the *unrolling factor*:

```
                                     for (i=1; i<n-1; i+=2) {
                                        a[i] = a[i]+a[i-1]*a[i+1];
for (i=1; i<n; i++)                     a[i+1] = a[i+1]+a[i]*a[i+2]; }
   a[i] = a[i]+a[i-1]*a[i+1];   →    if ((n-1)%2 != 0)
                                        a[n-1] = a[n-1]+a[n-2]*a[n];
```

It is a fundamental technique for generating the long instruction sequences required by VLIW machines (*very long instruction word*). This technique helps in increasing instruction-level parallelism if both assignments from the code can be executed simultaneously while updating the loop variables. Locality can be improved if array elements used in the unrolled loop are kept in registers.

Loop tiling is primarily used to improve cache reuse by dividing an iteration space into tiles and transforming the loop nest to iterate over them:

```
                                  for (ti=0; ti<n; ti+=64)
for (i=0; i<n; i++)                  for (tj=0; tj<n; tj+=64)
   for (j=0; j<n; j++)      →           for (i=ti; i<min(ti+64,n); i++)
      a[i][j] = b[j][i];                   for (j=tj; j<min(tj+64,n); j++)
                                              a[i][j] = b[j][i];
```

In the example above, a is assigned the transpose of b. By iterating over sub-rectangles of the iteration space after loop tiling, the loop uses every cache line fully.

The issue of finding an optimal sequence of loop optimizations has been discussed in [Wolf and Lam, 1991]. In their work, loop transformations are represented uniformly using integer matrices. Dependencies between iterations of a loop nest are modeled by means of distance vectors. This way, a problem such as maximizing parallelism can be reduced to the determination of a unimodular transformation matrix that maximizes the objective function. Efficient algorithms for the calculation of a transformation matrix are presented.

2.6.2   Code Generation for embedded Processors

The generation of optimized machine code for different architectures of embedded processors is treated in detail by [Marwedel and Goossens, 1995] and [Leupers, 2000]. Here, the code generation process is divided into different phases (compare section 1.1.2) which are solved sequentially. The authors presents several "close-to-silicon" techniques dealing with code selection, register allocation and scheduling for different classes of embedded processors.

For multimedia processors, a code selection technique exploring SIMD instructions (*single instruction, multiple data*) is introduced. These instructions perform a set of identical computations on sub-registers in parallel and are supported by various common processors, e. g. Intel Pentium MMX [Intel Corp., 2002] or Philips TriMedia [Philips Corp., 1997]. Using a sophisticated tree pattern matching approach combined with integer linear programming for code selection, a high utilization of SIMD instructions has been achieved.

Average runtime speed-ups of 7% are reported by Leupers after generating code making use of *conditional instructions*. Processors supporting this concept allow to attach a boolean guard expression to every regular machine instruction. When control flow reaches an instruction, the value of the attached guard decides whether to execute the instruction or not. This kind of low-level control flow optimization helps in reducing costly explicit conditional jump instructions. Due to the fact that only the best implementation for every *if*-statement occurring in a program is chosen (i. e. using either conditional instructions or conventional conditional jumps), the relatively small speed-ups are not surprising due to the local scope of this kind of optimization. In contrast, chapter 5 shows that much higher gains can be achieved by minimizing the total number of *if*-statements in a program using high-level global control-flow analysis techniques.

Code generation techniques based on the paradigm of *phase coupling* have been published by [Lorenz et al., 2001] and [Bashford and Leupers, 1999]. The key idea of this concept is to solve the problems of code selection, register allocation and scheduling simultaneously. This way, the interdependencies of these phases can be modeled accurately leading to better code compared to the isolated execution of the phases. Lorenz makes use of genetic algorithms in order to model the code generation problem, whereas constraint logic programming is employed by Bashford.

Chapter 3

FUNDAMENTAL CONCEPTS FOR OPTIMIZA-TION AND EVALUATION

This chapter provides a brief introduction into different kinds of techniques which are commonly used in this book. The contents of this chapter is kept at the level of an overview, so that only those aspects are covered which are necessary for a full understanding of the source code optimization techniques presented in chapters 5 to 7. References to more detailed and advanced literature are given in each section of this chapter.

In section 3.1, geometric models called polytopes and related operations and algorithms are presented. These structures are used in this book in order to model control flow aspects of loop nests. Since various genetic algorithms are presented in chapter 5, an introduction into the terminology and fundamental concepts of genetic optimization is given in section 3.2. Finally, the compilation and simulation techniques employed for generating all kinds of experimental results given in this book are introduced in section 3.3.

3.1 Polyhedral Modeling

A polyhedron is a geometric representation of systems of linear equations and inequations. These models have been studied in several related areas of research: from the geometric point of view by computational geometrists, from the algebraic point of view by the linear programming and operations researchers, and from the matrix point of view by the high-performance computing community. Formally, a polyhedron is defined as follows:

DEFINITION 3.1 (POLYHEDRON & POLYTOPE)

a A set $P = \{ x \in \mathbb{Z}^N \mid Ax = a, \ Bx \geq b \}$ is called a **polyhedron** for matrices $A, B \in \mathbb{Z}^{m \times N}$ and vectors $a, b \in \mathbb{Z}^N$ and $N, m \in \mathbb{N}$.

b A polyhedron P is called a **polytope** if $|P| < \infty$.

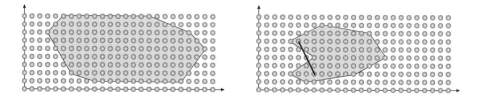

Figure 3.1a.   A convex Polytope   *Figure 3.1b.*   A non-convex Set of Points

Informally, a polyhedron is a subspace of $\mathbb{Z}^N$ which is bounded by a finite number of linear constraints.

DEFINITION 3.2 (CONVEXITY)
*A subspace $S \subset \mathbb{Z}^N$ is said to be **convex** if for each pair of points $x, y \in S$, every convex combination $ax + by$ of x and y with $a + b = 1$, $a, b \geq 0$, is also included in set S.*

A set S is convex if every line segment with endpoints in S is also included in S. According to definition 3.2, it was shown that polyhedra necessarily are convex sets [Wilde, 1993]. This fundamental property of polyhedra is illustrated in figure 3.1a. The shaded region depicted in this figure represents a polytope, since every line starting and ending in this area only contains points that are also within the region. In contrast, the region shown in figure 3.1b is not a polytope, because the plotted line crosses a point outside this zone.

The convexity of a set $S \in \mathbb{Z}^N$ is important because it provides a criterion for the closure of operations defined on sets.

DEFINITION 3.3 (OPERATIONS ON SETS)
Let $A, B \subset \mathbb{Z}^N$.

a *The **intersection** of A and B is defined by*
  $A \cap B := \{\, x \in \mathbb{Z}^N \mid x \in A \wedge x \in B \,\}$

b *The **union** of A and B is defined by*
  $A \cup B := \{\, x \in \mathbb{Z}^N \mid x \in A \vee x \in B \,\}$

Since polyhedra are subsets of $\mathbb{Z}^N$, the intersection and union operations as specified in definition 3.3 can also be applied to polyhedra. Due to the fact that the intersection of convex sets is still convex, polyhedra are closed under the intersection. In contrast, the union operation is not closed for polyhedra, because the resulting sets are not necessarily convex (see [Wilde, 1993]).

The union is a basic operation on sets which is also required for the optimizations presented in this book. In order to overcome the limitations of this operation, finite unions of polyhedra originally presented in [Wilde, 1993] are used in the following:

DEFINITION 3.4 (FINITE UNION OF POLYHEDRA)
*For polyhedra P_1, P_2, ..., P_n and $n \in \mathbb{N}$, $n < \infty$, $U := \{P_1 \cup P_2 \cup \ldots \cup P_n\}$ is a **finite union of polyhedra**.*

From the algorithmical point of view, finite unions of polyhedra are nothing but a linear list of polyhedra which are implicitly connected using the union operator. For these structures, the intersection and union operations are defined as follows:

DEFINITION 3.5 (OPERATIONS ON FINITE UNIONS OF POLYHEDRA)
Let $A = \bigcup_i A_i$ and $B = \bigcup_j B_j$ be finite unions of polyhedra.

a   *The **intersection** of A and B is defined by*
$$A \cap B := (\textstyle\bigcup_i A_i) \cap (\bigcup_j B_j) = \bigcup_{i,j}(A_i \cap B_j)$$

b   *The **union** of A and B is defined by[1]*
$$A \cup B := (\textstyle\bigcup_i A_i) \cup (\bigcup_j B_j)$$

In [Wilde, 1993] it has been shown that finite unions of polyhedra are closed under the operators given in definition 3.5. An undesirable side-effect of the intersection operator is the rapid size increase of the resulting polytope description due to the coalescing of the two constraint sets (compare definition 3.5a). For this reason, minimizing the set of constraints has proven to be a necessity in order to restrict the computational effort of polyhedral manipulations.

The actual technique employed for minimizing the set of linear constraints is based on the *double description of polyhedra*. In 1896, Hermann Minkowski [Minkowski, 1896] proved that any polyhedron P defined implicitly by a set of linear equations and inequations (cf. definition 3.1) has an equivalent dual representation. This dual representation is characterized by a finite set S of points whose convex hull is P. The convex hull of S consists of the convex combinations of all points of S and is thus the smallest convex set containing S.

An algorithm to compute the dual representation of a polyhedron P has been proposed in [Motzkin et al., 1953]. This method starts with an initial set of points S. During each iteration of Motzkin's algorithm, one new constraint of P is considered. At each step, a new set of points S' is computed by modifying the set S of the previous iteration. This is done in such a way that S' always is a minimal set of points whose convex hull represents all constraints of P

---

[1]From the algorithmical point of view again, the union of A and B is realized by a simple concatenation of the linear lists representing A and B.

which have already been processed by the algorithm. The complexity of this technique is $\mathcal{O}(n^{\lfloor \frac{d}{2} \rfloor})$, where n is the number of constraints of P and d is its dimension. Due to the fact that Motzkin's algorithm only results in a minimal set of points S for a polyhedron P, P still can contain redundant constraints and thus be non-minimal. As a consequence, this algorithm has been extended by Wilde to minimize the set of constraints of P. Basically, this is achieved by eliminating redundant constraints and by performing Gaussian eliminations to solve the system of equalities and to eliminate as many variables as possible.

One of the fundamental problems in the context of polytopes is the counting of integer points, i. e. points contained in a polytope having integer coordinates. In the domain of code optimization for high-performance computing, this value often models the number of memory locations touched by a loop or the iteration space of a loop nest.

This problem is a typical instance of an *enumeration problem*. In contrast to a *search problem* where only one valid solution for an instance of the problem has to be found, the total number of all solutions for a problem instance is subject of this problem class. Enumeration problems associated with *NP*-complete problems are clearly *NP*-hard [Garey and Johnson, 1979]. Furthermore, even in the unlikely case that $P = NP$ holds, it is probable that such enumeration problems remain intractable in polynomial time. Since the search problem associated with the size of polytopes deals with the question if a point exists satisfying all constraints and is *NP*-complete, the computation of a polytope's size is even harder. Actually, it has been shown that this problem is #*P*-complete [Kaibel and Pfetsch, 2003].

In [Clauss and Loechner, 1998], methods are developed for deriving a closed-form symbolic formula for the number of integer points contained in a polytope. If the set of integer points to be counted lies inside a union of convex polytopes, the number of points can be expressed by a special kind of polynomial called *Ehrhart polynomial*. Since the computation of these polynomials requires the translation of a polytope into its dual representation, the algorithms presented in [Clauss and Loechner, 1998] for polytope size calculation also have exponential worst-case complexity.

Although several tasks dealing with polytopes can have exponential runtimes in the worst-case, the utilization of the methods described in this section has proven to be highly efficient in the context of source code optimization (see the experimental results given in this book). The algorithms used effectively for the automation of the source code transformations presented in the following chapters stem from the polyhedral library developed at IRISA, France [Loechner, 1999].

For this library, a tool called *CLooG* [Bastoul, 2003] exists which is able to produce C code according to a given finite union of polytopes. However, CLooG is not used in the context of the source code optimizations presented

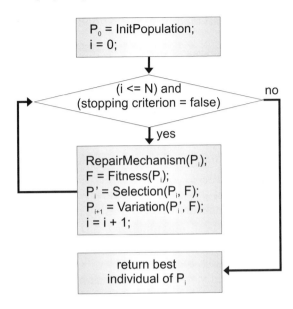

Figure 3.2. Schematic Overview of Genetic Algorithms

in this book. This is due to the fact that CLooG only generates C pseudo-code consisting of loops and *if*-statements visiting all integral points of a finite union of polytopes in their lexicographic ordering. Since the code created by CLooG only contains information about the control flow, all program code responsible for data processing is not included making the full implementation of source code optimizations more complicated.

3.2 Optimization using Genetic Algorithms

Genetic algorithms have proven to solve complex optimization problems by imitating the natural optimization process [Bäck, 1996, Holland, 1992]. In nature, a steady overproduction of creatures can be observed, although only a limited amount of resources like nutrition or habitat is available. The resulting competition allows only those individuals to survive which are best adapted to their environment and which are able to stand up to their rivals (*"survival of the fittest"*). As a consequence, the fitness of an individual is the crucial criterion for the procreation of descendants and for passing the genetic information to the next generation of individuals.

Genetic algorithms are probabilistic methods which simulate the natural evolution process by generating a number of random solutions. Some of these solutions are selected and may pass their genetic information to the next generation and are variated randomly afterwards. Genetic algorithms iteratively improve the solution by applying small modifications to good solutions to obtain better solutions.

An overview of the main steps of the optimization process of genetic algorithms is given in figure 3.2.

In a first step, a set of solutions (*individuals*) is generated. This set of individuals is called a *population*. The *initial population* P_0 represents the first *generation* during the evolutionary process. Then, the evolutionary algorithm iteratively computes new populations until a maximum number N of generations has been reached or a *stopping criterion* is fulfilled. For each generation i, all individuals of the population are examined. In some cases, the individuals need to be manipulated by a *repair mechanism* in order to obtain valid solutions. Then, all individuals are evaluated using a *fitness function F*. During the *selection* phase, the best individuals $P_i' \subset P_i$ from population P_i are selected according to their *fitness*. A new population P_{i+1} is computed, varying the individuals of P_i' using genetic operators like crossover and mutation in the *variation phase*. Finally, after all iterations have been executed or the stopping criterion has been fulfilled, the best individual found during the entire optimization process represents the final solution.

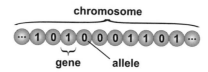

Figure 3.3.   Genetic Algorithm Encoding

Figure 3.3 depicts the way of encoding solutions using genetic algorithms. Every individual of a population is represented as a *chromosome* consisting of a sequence of *genes*. The actual value which is encoded by a gene is called *allele*. In many approaches, these alleles are binary values as shown in the figure above. Though, genes are also allowed to represent discrete values in general.

Genetic operators vary given chromosomes to produce new modified solutions. The variation step of genetic algorithms generally consists of the application of two basic genetic operators to individuals as illustrated in the following figures.

a The *crossover* operator probabilistically recombines the genes of two chromosomes. In figure 3.4a, this is illustrated by means of the so called *single point crossover*. In this case, two chromosomes are split into two parts which are then assembled vice versa. Other variants of the crossover operator have been studied in literature [Drechsler, 1998], e. g. the two point crossover where two points (two vertical lines) are probabilistically defined

Figure 3.4a.   Crossover   Figure 3.4b.   Mutation

for the chromosomes and only the genes between these two points are interchanged.

b The *mutation* operator changes some probabilistically chosen genes of a chromosome by replacing its contents with another allele as shown in figure 3.4b.

On the one hand, the progress during genetic optimization is achieved by the genetic operators which generate entirely new individuals or just slightly modify existing ones. On the other hand, the selection mechanism is responsible for the creation of new generations of individuals. As can be seen, the fitness function is the key component of every genetic algorithm since it measures the quality of each individual and thus directly influences the selection process. The fitness function has to be designed carefully in order to ensure that the genetic optimization process finally leads to high-quality solutions. In particular, the following aspects for fitness functions have to be considered:

- In general, the fitness function is a mapping $F : I \rightarrow \mathbb{R}$ computing a real number for an individual of a population. The fitness function should have a fine granularity so that only few different individuals are mapped to the same fitness value. Ideally, the fitness function should be an injective mapping.

- The fitness function should be a "monotonic" function in terms of the quality of individuals. Situations have to be avoided where the fitness of individual I is lower than the one of I', although individual I represents a better solution for the underlying optimization problem.

- Particular demands on the fitness function arise if the genetic algorithm does not use a repair mechanism as previously mentioned. Here, a population P_i can contain invalid individuals which do not represent a valid solution for the optimization problem.

  In such a situation, it has to be ensured that any valid individual always has a higher fitness than an invalid solution. Additionally, the fitness function should compute meaningful values also for invalid individuals so that a better individual I has a higher fitness than I' even if I and I' both are invalid. These properties guide the genetic algorithm towards the space of valid solutions, since hopefully the fitness function helps to derive a valid individual out of invalid ones during the subsequent generations.

The execution of a genetic algorithm and of its genetic operators depends on several parameters. The value N denoting the maximum number of generations has already been mentioned. The most important remaining possibilities for adjustments are:

- The *population size* denotes the number of individuals contained in every population P_i.

- The *replacement fraction* is an important parameter for the selection of individuals. It defines the percentage of individuals of a population which are not allowed to pass their genetic information to the next generation and thus are replaced by other individuals. This way, the size of the set P_i' shown in figure 3.2 is determined.

- For the genetic operators, separate values representing the *crossover* and *mutation probability* can be given. These values specify how often crossover and mutation should be used in order to create new or modified individuals for the next generation.

The advantages of genetic algorithms are manifold:

- Due to the random variation of individuals and the selection of the fittest individuals, genetic algorithms are able to revise unfavorable decisions made during earlier generations so that local extrema can be overcome. As a consequence, genetic algorithms are adequate for solving non-linear optimization problems.

- Genetic algorithms can be applied to *NP*-hard or *NP*-complete problems where the use of known exact methods is not practicable due to the high computation times.

- Since the maximum number N of generations is defined before the execution of a genetic algorithm, the computation time of genetic algorithms is predictable.

The disadvantage of genetic algorithms and all other heuristic algorithms is the potential loss of optimality. Using genetic algorithms in order to solve an optimization problem, it can not be guaranteed that the best solution found during genetic optimization is the optimal one. In contrast to branch and bound heuristics (e. g. integer linear programming), it is even impossible to estimate the closeness of an individual to the optimal solution. In addition, the repeated application of a genetic algorithm to the same instance of an optimization problem can lead to varying best solutions from one run of the algorithm to the other, depending on the behavior of the random number generator used during mutation and crossover.

Nevertheless, genetic algorithms often lead to solutions having a very high or near-optimal quality. For these reasons, genetic algorithms have been utilized successfully in the domains of code generation for irregular architectures [Cooper et al., 1999, Leupers and David, 1998, Lorenz and Marwedel, 2004], hardware synthesis [Landwehr, 1999] and codesign [Niemann, 1998].

In the context of the source code optimization techniques presented in this book, all genetic algorithms are implemented using the PGAPack Parallel Genetic Algorithm library [Levine, 1996]. This library provides the user with a full set of data structures and algorithms realizing all components and operators for genetic algorithms described in this chapter. The use of this library for genetic algorithms is advantageous, because all details concerning the application flow of genetic algorithms shown in figure 3.2 are hidden. This way, the programmer can concentrate on the formulation of an appropriate genetic encoding and fitness function.

All algorithms provided by PGAPack are highly parameterizable. By default, the following parameters are used by the genetic library:

- Population size: 100

- Maximum iterations: 1,000

- Replacement fraction: 10%

- Crossover probability: 0.85

- Mutation probability: 1 / *chromosome length*

These default values are used by all genetic algorithms described in this book. For several reasons, the evaluation of different combinations of parameters in order to improve the quality or the runtimes of the algorithms is omitted here. First, all solutions generated by the genetic algorithms have a very high quality as can be seen from the experimental results given in this book. Second, all genetic algorithms presented in the following are highly robust in the sense that manifold application of the algorithms to the benchmarks used in this book always leads to the same solutions. No variations between different runs of the algorithms were observed. Third, the runtimes of all genetic algorithms are very low. The genetic optimization processes terminate after a few CPU seconds. Due to these reasons, no necessity for the modification of parameters is given.

Since all fundamental algorithms and parameters are given by the PGAPack library, the descriptions of the genetic algorithms in this book are reduced to formal definitions of the chromosomal encodings and the developed fitness functions.

3.3   Benchmarking Methodology

Because of the fact that all optimizations presented in this book are source code transformations based on the standardized ANSI-C programming language [Kernighan and Ritchie, 1988], the techniques described in the following chapters are highly retargetable. All optimized C source codes are accepted and compiled by any ANSI compatible C compiler regardless of the actual processor for which the compilation is initiated. This processor independence allows a very detailed benchmarking to be performed on a large variety of different programmable processors. Since the processors considered in this book represent different classes of CPUs (DSP, VLIW, embedded RISC, ...), the effects of all optimizations on these different CPU classes can be quantified with a very high degree of accuracy.

Depending on the individual source code transformations, different objectives for optimization can be pursued. The primary goal of all optimizations presented in this book is to improve the runtime performance of typical benchmarks. In order to identify in detail the concrete factors leading to a measured change of runtimes, a methodology used to quantify internal CPU events like pipeline and cache behavior is described in section 3.3.1. Often, the acceleration of a benchmark is associated with increases in code size. Therefore, the method for measuring runtimes and code sizes during benchmarking is described in section 3.3.2. Finally, the design flow employed in order to measure the energy dissipation of a program is briefly presented in section 3.3.3.

More technical information about the used compilers and optimization flags as well as detailed lists of the data gathered during benchmarking can be found in the appendices.

3.3.1   Profiling of Pipeline and Cache Performance

Modern processors often contain dedicated hardware for performance monitoring. This allows a programmer to have a close look into the processor during the execution of a program. This way, the behavior of a program or only of some of its core algorithms can be monitored for a particular architecture so that areas of performance losses in the code can be determined and stall conditions can be remedied.

In order to identify stall conditions, such processors are equipped with particular configurable counters allowing the user to gather information about the performance of applications by keeping track of events during code execution. These counters are configurable so that the programmer can specify the system events to be monitored by the single counters. Depending on the processor, medium to large sets of different events can be traced, often including the following features:

- General CPU behavior (e. g. clock cycles, executed instructions)

- Pipeline behavior (e. g. pipeline stalls, decoded instructions)

- Branch prediction performance (e. g. taken branches, mispredicted branches)

- Memory system performance (e. g. accesses and misses of data and instruction caches / TLBs, main memory accesses)

In the context of the source code optimizations presented in this book, performance measuring hardware is used for the evaluation of the extent to which a source code transformation leads to improvements of the internal CPU behavior. For this purpose, a benchmark is compiled before and after the application of a source code transformation as described in the following section 3.3.2. By invoking the compiled programs and configuring and enabling the performance measuring support, highly reliable results about the effects of an optimization can be achieved. In this work, the focus lies especially on pipeline and cache performance, since typical embedded processors are often pipelined (e. g. TI C6x [Texas Instruments Inc., 1999], ARM7TDMI [ARM Ltd., 2001]) and use several caches [Catthoor et al., 2002] (in the absence of hard real-time constraints). It has turned out that these factors frequently impose a major bottleneck to the performance of embedded software.

Since not all the processors listed in section 3.3.2 include performance counters, detailed results based on measurements of system events are only given for the following CPUs in the subsequent chapters:

- Sun UltraSPARC III

- Intel Pentium Pro MMX

- MIPS R10000

This profiling based methodology for performance evaluation is especially useful in the case of pipeline and cache performance, because it is commonly accepted that simulation software which has to model complex instruction pipelines and memory hierarchies often is inaccurate.

3.3.2 Compilation for Runtime and Code Size Measurement

Measuring the runtime of a program given as an ANSI-C source code is a simple task, because the source codes only need to be compiled and executed.

Due to this simplicity and since C compilers are available for a large set of processors, runtime measurements were performed on a large range of different CPUs in order to obtain the experimental results presented in this book. Actually, the following processors including embedded RISC architectures, VLIW machines, DSPs and RISC workstations were used:

- Sun UltraSPARC III

- Intel Pentium Pro MMX

- HP-PA 9000

- MIPS R10000

- Power-PC G4

- DEC Alpha EV4

- Philips TriMedia TM-1000

- Texas Instruments TMS320C62

- ARM7TDMI[2]

Runtimes are measured by executing and profiling the compiled programs on existing hardware using either available workstations or evaluation boards. This approach has the advantage that measurements can be done at a very high speed without depending on slow and often imprecise simulators. In order to obtain reliable values for processors integrated in a multitasking environment, all interfering background processes have been disabled. Additionally, a total of twenty runtime measurements was performed in such a case. The slowest and the fastest result were discarded and the average of the remaining eighteen runtimes was used.

Usually, the notion of code size refers to the total amount of memory required for storing and executing a program on a given platform. This includes the sizes of all executable parts of a program (also possibly required startup code) as well as the size of global data, arrays, etc. In contrast to this common definition, code size in the context of this book only focuses on the static size of the executed program. The reason is that the source code optimizations presented in the following only transform the program code without touching the associated data and its layout. In order to quantify the changes of code sizes after source code transformation, only the instructions of the program code are considered.

The code sizes of the benchmarks are determined by compilation of the C source codes such that assembly listings are generated instead of executable

---

[2]The ARM processor is considered in its 16-bit THUMB-mode as well in the 32-bit ARM-mode.

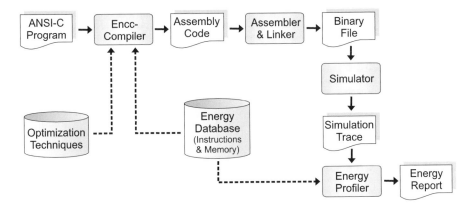

Figure 3.5. Overview of the Workflow employed in the *encc* Compiler

binaries. From these assembly listings, all comments, labels, symbol and debugging information etc. was stripped so that only the assembly instructions
remained. The total amount of assembly mnemonics occurring in these files is
then taken as a measure for the code sizes.

Whenever a compiler is invoked in order to translate a C source code, this
is done using the highest degree of optimization provided by the compiler.
Only when using all available compiler optimizations, a fair judgment about
the quality of the source code optimizations presented in this book can be given.
A compilation without using compiler optimizations is not meaningful, because
measured improvements can vanish if a compiler optimization performs a similar transformation as the one applied at the source code level. Consequently,
only measurements based on highly optimized compiled codes are substantial
since improvements measured this way clearly indicate that the applied source
code transformation goes beyond the techniques included in today's state-of-
the-art optimizing compilers.

3.3.3   Estimation of Energy Dissipation

Since power efficiency has become a crucial issue for a broad class of embedded applications, it is important to evaluate the savings of energy dissipation
achieved by the techniques presented in the following chapters. For this purpose, a compiler framework for the energy aware compilation of C programs
for the ARM7TDMI processor is used. An overview of the workflow of this
framework is given in figure 3.5.

The *encc (energy aware C compiler)* [Steinke et al., 2002a] is an ANSI compatible C compiler whose goal is the generation of code with minimized energy

Memory Type	Size	Energy [nJ]	Access Time [Cycles]
Scratchpad	128 B	0.53	1
Scratchpad	256 B	0.61	1
Scratchpad	512 B	0.69	1
Scratchpad	1 kB	0.82	1
Scratchpad	2 kB	1.07	1
Scratchpad	4 kB	1.21	1
Scratchpad	8 kB	2.07	1
Main Memory (16 bit)	512 kB	24.0	2
Main Memory (32 bit)	512 kB	49.3	4

Table 3.1. Examples of Data provided by the Energy Database

consumption. In order to achieve this, *encc* makes use of a large variety of compiler optimizations. These include well-known techniques described in standard
literature on compiler design [Muchnick, 1997], but also particular optimizations for energy consumption reduction like e. g. register pipelining [Steinke
et al., 2001b] or energy aware code selection [Steinke, 2003]. For the minimization of power consumption, *encc* uses an energy database providing the
energy consumption per executed instruction and per memory access. This
energy model [Steinke et al., 2001a] is based on physical measurements using
available hardware and has a precision of 1.7%. Some examples of the data
stored in the energy database are given in table 3.1.

The assembly code generated by the *encc* compiler is fed into an assembler
and linker so that an executable program is built. Since it is not practicable
to measure the total energy consumption of this binary program physically
using hardware, a simulation based approach is employed instead. During the
simulation, all executed instructions and accesses to different kinds of memory
are logged in a trace file which is finally processed by a tool for energy profiling.
The energy profiler combines this data with the individual energy consumptions
taken from the energy database and sums up the resulting values finally leading
to the total energy consumption of the executable program.

3.4   Summary

In this chapter, a brief introduction into a series of different techniques is
provided which is essential for the full understanding of the source code optimizations presented in the following chapters 5 to 7. First, the notion of
polyhedra and polytopes is defined. Some comments on the difficulties in ma-

nipulating these structures are given as well as mentioning the complexity of some algorithms related to polyhedra. Second, an overview of the terminology and function of genetic algorithms is presented. A genetic algorithm library is referenced and a set of parameters steering this library are listed. This way, the description of genetic algorithms in this book can be reduced to the definition of the chromosomal encoding and the fitness function. Finally, the different methodologies used to generate the experimental results presented in this book are explained and motivated.

Chapter 4

INTERMEDIATE REPRESENTATIONS AND THEIR SUITABILITY FOR SOURCE CODE OPTIMIZATION

For every automated approach of code generation for programmable processors, appropriate *intermediate representations* are required. An intermediate representation is used to model all aspects of a program that is intended to be executed on a processor. This includes all instructions of the program as well as its symbolic information like variables and data types. Using such a well-defined formal model, algorithms for program analysis and optimization can be applied automatically in order to generate code for a program (compare the traditional compilation process described in section 1.1.2). It is the goal of this chapter to give an overview of various intermediate representations and their place in the entire code generation and optimization process. Additionally, the choice of an intermediate model selected for optimization at the source code level is exposed.

Depending on the code generation algorithms and the considered programmable processor, the overall structure of an intermediate representation has to be chosen carefully. Intermediate representations can generally be divided into the following categories (see [Muchnick, 1997]):

- **Low-level intermediate representations:**
  Low-level intermediate representations frequently correspond almost one-to-one to target-machine instructions and are often strongly processor architecture dependent. The use of such models allows maximal optimization for a processor to be performed on the intermediate code in the final phases of compilation. Due to the closeness of the intermediate format to the machine instructions, the generation of assembly code is a relatively simple task.

- **Medium-level intermediate representations:**
  Medium-level intermediate formats are generally designed to reflect the range of features of source languages, but in a language-independent way,

41

and are designed to be good bases for generation of efficient machine code for one or more architectures. They provide a way to represent source variables, temporaries and registers. Control flow is reduced to simple conditional and unconditional branches and returns. All operations necessary to support block structures and procedures are made explicit.

Medium-level intermediate representations are appropriate for most of the optimizations done in compilers, such as common subexpression elimination or code motion.

- **High-level intermediate representations:**
  High-level intermediate representations are mostly used in the earliest stages of the compilation process. Their main characteristic is their closeness to the source language so that intermediate code can be transformed back into source code easily.

  One frequently occurring form of a high-level intermediate format is the *abstract syntax tree* which makes the structure of a program explicit. The tree, along with a symbol table indicating the types of the variables, provides all the information necessary to reconstruct the original source code. Alternatively, a high-level intermediate representation can also be organized in a linear instead of a tree-like form where the instructions of a program are stored in the same sequence as occurring in the source code. In either case, high-level features of the source language, such as array subscripts and loop structures, are essentially preserved in their source forms.

In the following sections 4.1 to 4.3, some popular intermediate representations are briefly presented according to the abstraction levels described above. In section 4.4, the choice of the intermediate format used for the implementation of the source code optimizations presented in this book is justified. For this purpose, an experimental comparison between two candidate models is provided.

4.1   Low-level Intermediate Representations
4.1.1   GNU RTL

The *GNU Compiler Collection* (GCC) [GCC, 2003] is a very popular family of C and C++ compilers. This compiler has been ported to various different architectures, among which also typical embedded processors can be found (e. g. ARM, Texas Instruments TMS320C4x or Tensilica Xtensa). Since this compiler seems to be highly portable – or is even structured to support retargetability – and also focuses on embedded platforms, the question arises if the intermediate format of the GCC is suitable for source-level optimizations.

Most of the work of the GCC compiler is done on an intermediate representation called *register transfer language* (RTL) [Stallman, 2002]. The syntax of

RTL is closely related to Lisp lists where each list item represents an expression. RTL defines several categories of expressions:

- Unary or binary arithmetic operations

- Comparisons

- Bit-field operators

- Jump- and Call-instructions

- Objects like registers or memory locations

- Auto-increment addressing instructions and others

Immediately after invocation, GCC builds a parse tree for a source file which has to be translated into a list of RTL expressions in a next step. Even during this early translation step, constraints imposed by the actual architecture have to be considered. For this purpose, a machine description containing information about available registers and machine instructions has to be provided for the generation of an RTL model for a source code.

Usually, the generation of an intermediate model for a source program is machine independent, whereas the application of optimizations to this model depends heavily on an architecture. In the case of the GCC, it can be seen that the process of RTL generation is already machine dependent.

This tight relationship between RTL and a machine description makes the GCC intermediate format unsuitable for the implementation of source code transformations as described in this book. It is impossible to resolve the conflicts between the inherent portability of programs optimized at the source code level and the processor dependence of RTL. Even the developers of GCC warn of using RTL for other purposes than described in [Stallman, 2002]:

> People frequently have the idea of using RTL stored as text in a file as an interface between a language front end and the bulk of GCC. This idea is not feasible.
>
> GCC was designed to use RTL internally only. Correct RTL for a given program is very dependent on the particular target machine. And the RTL does not contain all the information about the program.

Due to these restrictions, the GCC framework based on RTL is not considered as an alternative for the realization of source code optimizations in this book.

4.1.2   Trimaran ELCOR IR

The *Trimaran* project [Trimaran, 2002], mainly developed at HP Research Labs, is an extensible compiler framework for research on exploitation of instruction-level parallelism. In this project, the focus lies on the code generator called *ELCOR*. This module performs instruction scheduling, register allocation and machine-dependent optimizations. For this purpose, it makes use of parameterizable architecture descriptions. Hence, it is evident that the formal models employed by ELCOR belong to the class of low-level intermediate representations.

In the ELCOR IR, a program consists of a graph of operations connected by edges. This operation graph represents both a traditional control flow graph and a data flow graph. The edges between operations model various kinds of dependencies:

- Control dependencies representing a sequential control flow

- Flow, anti and output dependencies on registers

- Data dependencies on memory locations

In order to perform different kinds of analyses, operations can be grouped into regions of code. Using this concept, code sequences with different entry and exit properties, loop bodies and compound regions can be represented. In addition, ELCOR IR includes mechanisms in order to model:

- Predicated execution of operations which can be guarded by a condition

- *Explicitly Parallel Instruction Computing* (EPIC) related information representing scheduling and machine resource usage within an operation

Since the Trimaran framework is a research platform, it does not provide a full tool chain for the translation of C programs into assembly code. Instead, the generated code represented by ELCOR IR is fed into a cycle-true simulation engine. This simulator is configurable by a machine description so that it can be retargeted to an actual processor architecture. The simulation results in runtime execution statistics giving information about execution time, branch frequencies and resource utilization. Due to this low abstraction level of Trimaran, the generation of C code out of ELCOR IR is not possible.

4.2   Medium-level Intermediate Representations

4.2.1   Sun IR

The intermediate representation used in the Sun compilers is called *Sun IR* [Muchnick, 1988, Muchnick, 1997]. It is a typical medium-level model, because parsers for different source languages (e. g. C, Fortran and Pascal) and code generators for different architectures (SPARC v8 or v9, x86) can be attached to Sun IR.

Due to its abstraction level, Sun IR does not contain any particular constructs derived from one of the programming languages or processor architectures. Its structure is kept simple so that a program is modeled by a linked list of elements representing executable operations and several tables representing declarative information. Executable operations are modeled using generic instructions for e. g. arithmetical or logical operators.

Analysis and optimization based on Sun IR is performed using the concept of *basic blocks*:

DEFINITION 4.1 (BASIC BLOCK)
*A **basic block** $B = (I_1, \ldots, I_n)$ is a sequence of instructions with maximal length, so that*

a *B can be entered only at the first instruction I_1 and*

b *B can be left only from the last instruction I_n.*

Informally, a basic block is a maximal instruction sequence with linear control flow. Only the last instruction of a basic block is allowed to be a jump instruction branching to some other part of the code. The control flow of a program is represented by a graph structure where each node represents a basic block. Directed edges are inserted between basic blocks B_i and B_j whenever the termination of B_i can directly lead to the execution of B_j.

In order to perform loop optimizations, the Sun compilers contain a loop discovery phase where the index variable, lower and upper loop bounds and loop strides are extracted from the control flow graph.

The two main reasons why Sun IR is not a practical solution for the implementation of source code transformations are:

- The fact that high-level control flow constructs like loops are broken down to jump instructions and labels is a disadvantage. Since this kind of information is particularly required for the loop transformations presented in this book (e. g. loop nest splitting, see chapter 5), the costly re-generation of this information out of the basic blocks is relatively cumbersome.

- Sun IR is a commercial proprietary product which is not available for research purposes. As a consequence, no infrastructure for the analysis, transformation and C code generation of Sun IR exists.

In general, Sun IR is powerful enough for being used for source code optimization. This is also underlined by the fact that the Sun compilers include several high-level loop optimizations. But due to the very strong encapsulation and hiding of this intermediate representation within the compilers, Sun IR is not considered in this book.

4.2.2   IR-C / LANCE

IR-C [Leupers et al., 2003] (and its functionally equivalent counterpart LANCE [Leupers, 2001] developed at ICD [ICD, 2003]) is designed as an executable intermediate representation for retargetable compilation and high-level code optimization. Since in IR-C, all instructions are represented in three ad-dress code format, all high-level C constructs such as loops, nested *if*-statements and address arithmetic are replaced by sequences of primitive statements.

For each given C source file, one IR-C file is generated which is structured into symbol tables and functions. Each function directly represents a function of the original C file and consists of a list of IR-C statements of different types:

- Assignments

- Jumps and conditional branches

- Labels

- Return statements

The main difference of IR-C compared to other typical medium-level inter-mediate representations is its executability. This means that IR-C is defined to be a subset of ANSI-C so that every IR-C file can be processed directly by a compiler. This property offers several opportunities for the use of IR-C. On the one hand, all algorithms based on IR-C can be easily validated by exploiting the executability of IR-C. For this purpose, an original C program and its IR-C representative have to be compiled and executed on a given platform. If the outputs of both programs differ, the validation has failed. On the other hand, the authors claim that every optimization performed on IR-C can be seen as a source code optimization.

Due to its similarity to assembly statements, code generators for differ-ent embedded architectures have been successfully implemented using IR-C / LANCE (e. g. Infineon NP [Wagner and Leupers, 2002], Systemonic Hiper-Sonic DSP [ICD, 2003]). Since these code generators base on tree pattern matching strategies [Aho et al., 1989], IR-C includes mechanisms in order to translate the three address instructions into data flow trees in C syntax. This step can be seen as a raise of the abstraction level of IR-C, because data flow trees are said to have a higher expressiveness than three address code [Leupers et al., 2003].

As already mentioned in the context of Sun IR, the lack of high-level control flow constructs is not necessarily a disadvantage of an intermediate represen-tation as long as analysis techniques exist which can re-compute this informa-tion. Due to the explicit focus of IR-C on platform-independence and high-level optimization, this intermediate representation is a potential candidate for the implementation of the techniques presented in the following chapters.

4.3 High-level Intermediate Representations

4.3.1 SUIF

The *Stanford University Intermediate Format* (SUIF) [Wilson et al., 1995] is a freely available framework focusing on research on optimizing compilers. It is delivered with parsers for ANSI-C and Fortran 77. The SUIF system is organized as a set of compiler passes built on top of a kernel that defines the intermediate format. All passes are implemented as separate programs that link with the SUIF kernel. Each pass typically performs a single analysis or transformation and then writes the results out to a file. SUIF files always use the same output format so that new passes can be freely inserted at any point in a compilation.

SUIF is available in two different releases called SUIF 1 and SUIF 2. These two releases differ fundamentally in the sense that the entire compiler infrastructure was redesigned during the transition from SUIF 1 to SUIF 2. Besides all technical differences among both releases, the basic concepts of the intermediate representation are the same. This holds especially for the high-level representation of SUIF described in the following. Since both releases of the SUIF intermediate representation base on similar concepts to a large extent, the remainder of this book explicitly focuses on SUIF 1 without considering SUIF 2 any more.

The representation of a program using SUIF includes both high-level and medium-level information. In the first stages of compilation, the high-level structure is represented by a language-independent form of abstract syntax trees. This format called *high-SUIF* is well suited for dependence analysis and loop transformations requiring the high-level structure of the code. In high-SUIF, loops, *if*-statements and array references are represented in the same way as they are used in the source languages. The abstract syntax trees of high-SUIF consist of nodes of the following kinds:

- *For*-loops composed of index variable, lower and upper bound, stride and loop body

- *Do-while*-loops composed of test condition and loop body

- *If*-statements composed of test condition and *then*- and *else*-part

- Blocks of code (i. e. all C statements between curly braces)

- Instructions

Later in the compilation process, the abstract syntax trees can be reduced to sequential lists of instructions called *low-SUIF*[1]. During this process, all high-

---

[1]Despite this denomination, low-SUIF in fact is a medium-level intermediate representation.

level constructs are dismantled to lower-level assembly-like statements. This way, an interface to a code generator for a given architecture (e. g. Machine SUIF [Smith and Holloway, 2002]) is provided by SUIF.

In the case that a SUIF compatible code generator is unavailable for a given processor, the SUIF intermediate representation can be exported as ANSI-C which can then be fed into any available compiler. During this translation of SUIF to C, all high-level constructs actually present in a SUIF file are directly mapped to the corresponding C statements, so that the generated C output also has a high abstraction level. This property makes SUIF very interesting for the realization of source code transformations.

4.3.2   IMPACT

The *Illinois Microarchitecture Project utilizing Advanced Compiler Technology* (IMPACT) [Hwu et al., 2003] is mainly an optimizing C frontend with emphasis on architectures supporting instruction-level parallelism. It includes C and Fortran parsers and code generators for various architectures (e. g. HP PA, SPARC, x86, TI C4x).

The intermediate representation is structured in two distinct layers called *Pcode* and *Lcode*. Pcode is the highest level of intermediate format based on a parallel C representation. In Pcode, the control flow is modeled using particular statements which represent (serial or parallelizable) *for*-loops or *if*- and *switch*-statements.

Pcode can be lowered to Lcode. Lcode is the lowest level of program intermediate representation and has the format of a generalized register transfer language. During the transition from Pcode to Lcode, the high-level constructs are removed from the intermediate representation. In Lcode, a function consists of a sequence of basic blocks. In each basic block, operations are described in three address notation. Since Lcode is machine independent, it is a typical medium-level intermediate representation.

In the entire flow of compilation provided by IMPACT, the generation of C source code out of the high-level Pcode data structures is not possible. In contrast, an approach for C code generation based on the medium-level Lcode has been proposed for emulation purposes [Olaniran, 1998]. As a consequence, the generated C code only has a very low abstraction level. Concluding, it should be noted that the IMPACT intermediate representation generally can be used for source level optimizations, though it is not as well suited as SUIF is.

4.4   Selection of an IR for Source Code Optimization

As can be seen from the previous sections, various intermediate representations with different characteristics exist. Before deciding which model is best suited for the automation of source code transformations, the demands on intermediate representations need to be defined:

- **Platform independence**
  One of the major advantages of optimizations at the source code level is the inherent portability of such techniques. If this property should not be lost, the involved intermediate representation has to be independent of concrete processor architectures.

- **Closeness to programming language**
  In order to perform loop analysis and optimization at source code level, it is useful when high-level control flow constructs like *for*-loops and *if*-statements are kept within the intermediate representation. If this is not the case, it has to be possible to recompute this kind of control flow information.

- **C code generation out of intermediate representation**
  It is not sufficient for an intermediate representation to be generated out of a source program and to model the programming language closely. In addition, it is mandatory to be able to export the intermediate models as program code, so as to establish a full source-to-source transformation framework.

- **Extensibility and modularity**
  The intermediate representation should have a modular structure and should be designed for being extended by auxiliary modules. This property helps in minimizing the effort required for the implementation of source code transformations. In this context, a file based structure, where the intermediate representation is written into a file after every optimization and read in by the next pass, is highly preferable.

Due to the lacking platform independence and their closeness to assembly code, low-level models can generally be excluded from the set of intermediate representations that are suitable for the implementation of source code optimizations.

Medium-level intermediate formats are basically well suited for this purpose. As stated in section 4.2, such models typically base on an assembly-like structure of control flow which is represented by conditional and unconditional jumps to labels. On the other hand, the formats presented in that section include mechanisms to determine loops out of the intermediate structure. Sun IR can be disqualified because of its strong encapsulation into the commercial Sun compiler and the non-present C code generation. In contrast, IR-C / LANCE explicitly bases on a C representation and thus fulfills all requirements listed above.

Intuitively, high-level intermediate representations appear to be the most appropriate structures for source level optimization. Nevertheless, IMPACT is

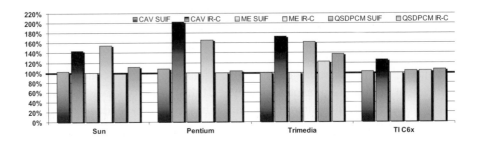

Figure 4.1. Experimental Comparison of Runtimes based on SUIF and IR-C/LANCE

excluded here, although it also satisfies all imposed demands. The reason for this can be seen in the very low abstraction level of the exported C code. If a transformation is performed based on high-level control flow constructs, the generation of medium- or low-level source code for compilation is disadvantageous. As an alternative, SUIF with its integrated support and export mechanisms of loops and *if*-statements is considered to be a suitable intermediate representation.

The above discussion shows that a fundamental decision between a medium-level intermediate representation or a high-level model has to be taken. For this purpose, an experimental comparison of IR-C/LANCE and SUIF was performed using the following benchmarks:

- CAV: Cavity detection algorithm in tomography images [Bister et al., 1989],

- ME: MPEG 4 full search motion estimation [Gupta et al., 2000],

- QSDPCM: Scene adaptive coding [Strobach, 1988]

In a first step, the source codes of these programs were translated into the high-SUIF and IR-C intermediate representations. The high-SUIF structures are exported back to C code in a second step. In order to obtain a fair comparison of high-SUIF and IR-C, the step of data flow tree generation is applied to the IR-C files. This way, C source codes based on SUIF and IR-C are generated for the benchmarks listed above. During this process, no kind of optimization was applied to the benchmarks. Optimization was completely left up to the compilers into which the source codes exported out of SUIF and IR-C were fed. Compilation and subsequent execution of the compiled codes were performed on a Sun UltraSPARC III, Pentium Pro, Philips TriMedia and TI C6x according to the benchmarking methodology explained in section 3.3.2.

Figure 4.1 depicts the runtimes of the benchmarks based on SUIF and IR-C on all processors (compare appendix A for the numerical data leading to this diagram). The runtimes are given relatively to the execution times of the original benchmark versions. These original runtimes are illustrated by the highlighted base line at 100%. For every benchmark and processor, the figure contains two bars representing the runtimes of the SUIF and the IR-C based program versions.

From figure 4.1 it can be seen that the runtimes of the SUIF based code versions are very stable compared to the performance of the original source codes. In almost all cases, the runtimes of the SUIF codes deviate only by up to 7.5% (CAV on Pentium) from the original ones. The only exception is the QSDPCM benchmark running on the TriMedia processor. In this situation, the SUIF code is 22.3% slower than the original program. These results clearly show that auto-generated code based on the SUIF intermediate representation stays very close to the original programs.

In contrast, the bars denoting the runtimes of the IR-C codes of the benchmarks are in most cases much higher than 100%. Only in four cases (ME on TI C6x and QSDPCM on Sun, Pentium and C6x), a code quality comparable to the one of SUIF was measured with a maximum degradation of 11.6% (QSDPCM Sun) compared to the original code. In all other cases, increases of runtimes between 26% (CAV on TI C6x) and even 103% (CAV Pentium) were observed. Concluding, it has to be noticed that the suitability of IR-C for source code optimization as claimed by [Leupers et al., 2003] is somewhat limited. Nevertheless, IR-C / LANCE is still very appropriate for the implementation of highly optimizing code generators.

These experimental results lead to the decision that a high-level intermediate representation is best suited for the purpose of source code optimization. Medium-level models are not preferable because compilers are obviously unable to generate efficient machine code out of three-address code lacking structured control flow. For a proper quantification of the effects of source code optimizations, a large influence of the used intermediate representation on the results is highly undesirable. In order to minimize this implicit influence, SUIF is used whenever an implemented algorithm for source code optimization is presented in the following chapters.

4.5 Summary

This chapter gave a condensed overview about different abstraction levels of intermediate representations for automatic code manipulation. Different actual intermediate representations and their key features were presented. This includes intermediate representations integrated in commonly used compilers as well as frameworks designed for research on code optimization. Finally, it was demonstrated experimentally that medium-level models should not be used in the context of source code optimization, even though they are in principle suited for this purpose. Instead, high-level representations should be preferred whenever auto-generated source codes are fed into external compilers for code generation.

Chapter 5

LOOP NEST SPLITTING

After the introduction of basic concepts and models in the previous chapters, the first source code optimization called *loop nest splitting* is presented in this chapter. The goal of this transformation is to generate a regular control flow in the innermost loops of data flow dominated software by minimizing the executions of *if*-statements. It has turned out that the effects of this optimization are manifold. Not only instruction pipeline stalls of embedded processors are reduced, but also runtimes, cache performance and energy dissipation are improved significantly.

In the following, the motivation for a control flow optimization performed on data flow dominated software is given in section 5.1. A survey of work related to loop optimizations and energy efficient data layout is provided in section 5.2. In section 5.3, the basic concepts of loop nest splitting and the employed analysis techniques are described. Section 5.4 contains extensions to these basic analysis algorithms enabling them to be applied to larger classes of embedded software. Detailed experimental results are presented in section 5.5, whereas a summary of this chapter is finally given in section 5.6.

5.1 Introduction

This section provides an introduction into different kinds of control flow overhead which can be found in the context of data flow dominated embedded software. First, it is shown in section 5.1.1 that applications of this kind (e. g. embedded multimedia algorithms) are usually programmed in such a way that *if*-statements are frequently executed during every loop iteration. Second, this behavior can be introduced by optimization techniques in order to exploit available memory hierarchies of an embedded system (see section 5.1.2). Finally, the way how loop nest splitting is applied to source codes and its different effects during runtime are explained in section 5.1.3 by means of a real-life example.

```
for (z=0; z<20; z++)
  for (x=0; x<36; x++) {
    x1=4*x;
    for (y=0; y<49; y++) {
      y1=4*y;
      for (k=0; k<9; k++) {
        x2=x1+k-4;
        for (l=0; l<9; l++) {
          y2=y1+l-4;
          for (i=0; i<4; i++) {
            x3=x1+i; x4=x2+i;
            for (j=0; j<4; j++) {
              y3=y1+j; y4=y2+j;
              if (x3 < 0 || 35 < x3 || y3 < 0 || 48 < y3)
                then_block_1;
              else
                else_block_1;
              if (x4 < 0 || 35 < x4 || y4 < 0 || 48 < y4)
                then_block_2;
              else
                else_block_2; }}}}}}
```

Figure 5.1.   A typical Loop Nest

5.1.1   Control Flow Overhead in Data Flow dominated Software

Data flow dominated applications such as medical image processing and video compression algorithms typically require large data memories. In order to manipulate its large quantities of data in an effective way, the program code of a data flow dominated application typically consists of deeply nested *for*-loops. With the help of the index variables of the loops, memory access pointers are calculated that are used for data manipulation. The main algorithmic part is usually located in the innermost loop. Very often, a multimedia algorithm has to treat particular parts of its data in a specialized way, e. g. pixels at the border of an image require a slightly modified algorithm than pixels situated in the center of an image. This boundary checking is implemented with *if*-statements in the innermost loop that check certain values of the index variables. A typical code fragment taken from an MPEG 4 full search motion estimation kernel [Gupta et al., 2000] written in ANSI-C is depicted in figure 5.1.

Despite already being written in an optimized manner (common subexpressions are eliminated and loop-invariant code is moved out of loops [Muchnick,

1997]), this code in its original structure has several properties that make it sub-optimal with respect to runtime and energy consumption. First, the *if*-statements in the innermost loop lead to a very irregular control flow. Every change in the linear control flow of a machine program, i. e. each jump instruction, causes a control hazard for pipelined processor architectures. This means that the pipeline needs to be stalled for a certain number of instruction cycles to prevent the execution of incorrectly prefetched instructions. Delay slots of a pipeline can sometimes be filled with useful instructions, but for deeply pipelined processors, this is not always the case. This leads to high jump penalties (Philips TriMedia TM1000: 3 instruction cycles / jump [Philips Corp., 1997]; TI C6201: 5 cycles / jump [Texas Instruments Inc., 1999]).

Second, the pipeline performance of a processor is also influenced by data accesses, since pipelines may have to be stalled during the execution of memory transfers. To realize the boundary checking mentioned above, the induction variables are accessed very frequently resulting in pipeline stalls if these variables can not be kept in processor registers. Degraded pipeline performance not only leads to increased execution times but also to a higher energy consumption due to the long idle times of the processor. Additionally, since it has been shown that 50% – 75% of the power consumption in embedded multimedia systems is caused by memory accesses [Stan and Burleson, 1995, Theokharidis, 2000, Wuytack et al., 1996], frequently repeated transfers of induction variables across memory hierarchies via system buses lead to a significant increase of energy consumption.

Finally, many instructions are required to evaluate the conditions of the *if*-statements. In the case of the motion estimation kernel shown above, these arithmetic and logical operations are in total as complex as the computations performed in the *then*- and *else*-blocks of the *if*-statements. This also demonstrates that execution time and energy consumption of such multimedia applications are affected adversely by *if*-statements in loop nests.

5.1.2 Control Flow Overhead caused by Data Partitioning

If the source code of an embedded application does not contain the control flow overhead described in the previous section, it is still possible that an energy aware optimizing compiler introduces several *if*-statements for control flow manipulation. At first glance, this behavior seems to be counterproductive, but it can be motivated by the efficient exploitation of memory hierarchies.

The emergence of portable or mobile computing and communication devices such as cellular phones, pagers, handheld video games etc. is probably the most important factor driving the need for low power design. Current battery technologies such as Lithium-Ion (Li-Io) have capacities of 90 Watt-hours / *kg* [Wahlström, 1999], meaning that ten hours of operation for a device that consumes 20*W* of operating power would require a battery weight of around

2.2kg. Thus, the cost and weight of batteries become bottlenecks that prevent the reduction of system cost and weight unless efficient low power design techniques are adopted.

As mentioned previously, the efficient utilization of memories is of major interest for the construction of low power devices. Usually large main memories are the kind of memories requiring most energy and clock cycles per access. On the one hand, the typically large amount of main memory accesses causes a high energy dissipation. On the other hand, the processor has to wait for several cycles during main memory accesses leading to an even higher energy consumption. To avoid these problems, a very effective way of energy reduction is to build up a memory hierarchy.

Additional memories are able to reduce the number of main memory accesses for frequently used instructions or variables. *Caches* are well known and included in many processor designs. Besides the data memory itself, they consist of two additional components [Hennessy and Patterson, 2003]. First, a tag memory is required to store information about valid addresses. Second, logic components are necessary for a fast comparison of addresses with the contents of the tag memory, so that cache hits and misses can be detected. The advantage of caches is their easy integration into a system since the detection of cache hits or misses is done automatically by the hardware. In the domain of embedded systems, caches are often not well suited due to their inherently high energy consumption during the accesses to the tag array and during tag comparison. Additionally, the difficult prediction of the cache behavior makes it often hard to determine accurate worst case execution times which is a crucial issue for embedded real-time systems.

Recently, the utilization of *scratchpad memories* has become an important alternative to caches [Kandemir et al., 2001, Banakar et al., 2002, Marwedel et al., 2004]. A scratchpad is a small memory mapped into the processor's address space requiring only simple address decoders. The absence of logic components checking the validity of data stored in the memory and the absence of parallel accesses to several sets are the reasons for the lower energy consumption of scratchpads compared to caches. However, these properties require a careful mapping of instructions or data to the memory which has to be done by the programmer or the compiler.

Optimizations for *data partitioning* (see [Verma et al., 2003]) have been presented where parts of a program and of its data are assigned in an optimal way (i.e. consuming least energy) to the scratchpad memory. Using *integer linear programming*, an instance of a knapsack problem is formulated for which the optimization of the objective function leads to the decision whether an array should be split, and at which point. The objective function is based upon an existing precise energy model [Steinke et al., 2001a] and models the energy savings which are achievable by data partitioning this way. The maximization

```
#define SIZE 100          #define SIZE 100
                          #define SPLIT 70
                          #define READ_ACCESS(value,index)
                             if (index < SPLIT)
                               value = Aleft[index];
                             else
                               value = Aright[index-SPLIT];

int A[SIZE];              int Aleft[SPLIT],Aright[SIZE-SPLIT];

for (i=0; i<SIZE/2; i++)  for (i=0; i<SIZE/2; i++)
  for (j=0; j<i; j++) {     for (j=0; j<i; j++) {
    data = A[i+j];            READ_ACCESS(data,i+j);
    ⋮                         ⋮
}                         }
```

Figure 5.2.   A typical Code Fragment before and after Data Partitioning

of this cost function by an ILP solver leads to the solution with the most energy efficient scratchpad usage. The benefit of this partitioning of arrays is that fragments of the original array can now be placed on the small scratchpad memory which was impossible before.

Since the compiler must guarantee that an application always accesses the correct part of a split array, *if*-statements dynamically selecting the appropriate sub-array have to be inserted into the program's code (see right half of figure 5.2). Although being beneficial with respect to energy consumption, this data partitioning step adds overhead to the embedded software leading to the same negative effects as previously described in section 5.1.1.

5.1.3   Splitting of Loop Nests for Control Flow Optimization

In this chapter, a new formalized method for the analysis and transformation of *if*-statements occurring in arbitrarily nested loops is presented solving a particular class of the *NP*-complete problem of the satisfiability of integer linear constraints. Considering the *if*-statements shown in figure 5.1 which are statically analyzable at compile-time, this method is able to detect that

- the conditions x3 < 0 and y3 < 0 of the first *if*-statement are never true,

- both *if*-statements are true for $x \geq 10$ or $y \geq 14$.

Information of the first type can be used to detect conditions not having any influence on the control flow of an application. This is a difference compared to conventional dead code, because dead code is defined as those instructions in a program that compute only values that are not used on any executable path leading from the instruction [Muchnick, 1997]. Since the boolean results computed by the conditions mentioned above are used within an *if*-statement, they will not be removed by a classical dead code elimination. Furthermore, a study of the assembly codes emitted by a large variety of state-of-the-art optimizing compilers shows that the compilers do not perform an induction variable analysis in order to determine that these conditions are unnecessary. Therefore, the removal of this kind of useless code using the proposed techniques leads to reductions in computational complexity and code size of a program and thus improves runtimes and energy consumption.

Using the information for which values the *if*-statements are true, the entire loop nest can be rewritten such that the total number of executed *if*-statements is minimized (see figure 5.3). To achieve this, a new *if*-statement (called the *splitting-if*) is inserted in the y-loop testing the condition x >= 10 || y >= 14. The *else*-part of this new *if*-statement is an exact copy of the body of the original y-loop. Since it is known that all *if*-statements are fulfilled when the splitting-if is true, the *then*-part consists of the body of the y-loop without any *if*-statements and their associated *else*-blocks.

To ensure that the splitting-if will not be evaluated repeatedly without need for values of $y \geq 14$, a second y-loop is inserted in the *then*-part of the splitting-if counting from the current value of y to the upper bound 48.

Example 5.1

> *Provided that variable x is equal to 7 and y is equal to 14, the splitting-if contained in the loop nest shown in figure 5.3 would evaluate to true due to the value of y. When assuming that the second y-loop would not be present in the code, the k-loop would be executed next. After its termination, y is incremented to 15, and the flow of control branches backward to the y-loop. Again, the splitting-if is satisfied for y equal to 15 so that the entire process repeats for all values of y up to 48.*

> *The presence of the second y-loop eliminates these repeated evaluations of the splitting-if. For y equal to 14, the splitting-if is executed once. Hereafter, the second y-loop is entered. During its execution, y is incremented up to 49 without any further evaluation of the splitting-if. For y equal to 49, the termination condition of this inner y-loop is met so that the flow of control is transferred back to the original first y-loop. Again, its termination condition is satisfied for y equal to 49 so that the next iteration of the x-loop for x equal to 8 would take place.*

It is important to see that this second y-loop modifies the index variable y which is also accessed by the original outer y-loop. Because of this fact, it is assured that the execution of both the new and the original y-loop leads to the assignment of exactly the same sequence of values to y as before the optimization. This way, the semantical correctness of the transformed program

```
for (z=0; z<20; z++)
  for (x=0; x<36; x++) {
    x1=4*x;
    for (y=0; y<49; y++)
      if (x >= 10 || y >= 14)              /* Splitting-If */
        for (; y<49; y++)                  /* Second y loop */
          for (k=0; k<9; k++)
            for (l=0; l<9; l++)
              for (i=0; i<4; i++)
                for (j=0; j<4; j++) {
                  then_block_1;
                  then_block_2; }
      else {
        y1=4*y;
        for (k=0; k<9; k++) {
          x2=x1+k-4;
          for (l=0; l<9; l++) {
            y2=y1+l-4;
            for (i=0; i<4; i++) {
              x3=x1+i; x4=x2+i;
              for (j=0; j<4; j++) {
                y3=y1+j; y4=y2+j;
                if (0 || 35 < x3 || 0 || 48 < y3)
                  then_block_1;
                else
                  else_block_1;
                if (x4 < 0 || 35 < x4 || y4 < 0 || 48 < y4)
                  then_block_2;
                else
                  else_block_2; }}}}}}
```

Figure 5.3.   Loop Nest after Splitting

is maintained. At the same time, these twofold y-loops lead to the minimization of executions of the splitting-if.

As can be seen from this example, the proposed optimization is able to generate linear control flow in the hot-spots of an application. Furthermore, references to memory are reduced significantly, because large amounts of branching, arithmetic and logical instructions and induction variable accesses are removed from the code. Using these factors, loop nest splitting is able to achieve con-

siderable speed-ups combined with reductions in energy consumption at the
expense of slightly increased code size.

5.2   Related Work

Loop restructuring transformations have been described in literature on compiler design for many years (see e. g. [Bacon et al., 1994, Muchnick, 1997]) and
are often integrated into state-of-the-art optimizing compilers. Conventional
loop splitting (also called *loop distribution* or *loop fission*) takes a loop that
contains multiple statements and splits it into two loops with the same iteration-
space traversal, such that the first loop contains some of the statements inside
the original loop and the second one contains the others as shown in figure 5.4.

```
for (i=0; i<n; i++) {              for (i=0; i<n; i++)
  a[i] = a[i]+c;                     a[i] = a[i]+c;
  x[i+1] = x[i]*7+x[i+1]+a[i];  →  for (i=0; i<n; i++)
}                                   x[i+1] = x[i]*7+x[i+1]+a[i];
```

Figure 5.4.   Conventional Loop Splitting

The main goal of this optimization is to create sub-loops with fewer data
dependencies allowing parts of an original loop to be executed in parallel [Ba-
con et al., 1994]. Moreover, instruction cache performance may be increased,
because locality is improved due to the smaller loop bodies. Additionally, the
smaller loop bodies can have the effect that code size constraints of a processor architecture with respect to zero-overhead hardware-loops [Uh et al., 1999]
are now met. As a consequence, control flow modifications after each loop
iteration can be performed by the processor hardware, making explicit branch
instructions in the code unnecessary. If hardware-loops can not be applied to
a loop, the control flow becomes more irregular after the application of loop
splitting, because a single loop is broken into many loops. In [Kandemir et al.,
2000] it is shown that loop splitting leads to increased energy consumption of
the processor core as well as the memory system. Also, computational complexity during loop execution is not reduced by this optimization, so that this
technique is not suitable for solving the problems discussed in sections 5.1.1
and 5.1.2.

Classical *loop unswitching* as depicted in figure 5.5 is applied when a loop
contains an *if*-statement with a loop-invariant test condition [Muchnick, 1997].
The loop is then replicated inside each branch of the conditional check saving
the overhead of conditional branching inside the loop, reducing the code size
of the loop bodies, and possibly enabling the parallelization of one branch of
the *if*-statement [Bacon et al., 1994]. The goals of loop unswitching and the
way how the optimization is performed on a loop are equivalent to the topics

```
for (i=0; i<n; i++) {
    a[i] = a[i]+c;
    if (x < 7)
        b[i] = a[i]*c[i];
    else
        b[i] = a[i-1]*b[i-1];
}
```
→
```
if (x < 7)
    for (i=0; i<n; i++) {
        a[i] = a[i]+c;
        b[i] = a[i]*c[i]; }
else
    for (i=0; i<n; i++) {
        a[i] = a[i]+c;
        b[i] = a[i-1]*b[i-1]; }
```

Figure 5.5. Conventional Loop Unswitching

mentioned in section 5.1. The main disadvantage of loop unswitching making it completely unsuitable for applying it to typical multimedia programs is that the *if*-statements must not depend on index variables. The main contribution of the transformation technique presented in this chapter is the explicit focus on loop-variant conditions for loop splitting. Since the required analysis techniques go far beyond those required for conventional loop splitting or unswitching and have to deal with entire loop nests and sets of index variables, this optimization technique is called *loop nest splitting*.

In the work of [Kim et al., 2000], the effect of several high-level compiler optimizations (loop unrolling, loop interchange, loop fusion and loop tiling) on memory system energy considering both instruction and data accesses are evaluated. The authors have observed that these techniques are effective in minimizing the energy consumed due to data accesses, but that the energy consumed due to instruction fetches is increased significantly. They draw the conclusion that techniques are required for the simultaneous optimization of the locality of data and instruction accesses. The experimental results given in this chapter show that loop nest splitting is able to achieve these goals.

The application of conventional loop splitting in conjunction with function call insertion was studied in [Liveris et al., 2002]. The impacts of these techniques on the instruction cache performance were quantified at the source code level using a formal model for instruction cache miss estimation. In their work, the authors also consider the importance of *if*-statements by using a probability for which conditions are satisfied. After the application of loop splitting, a large reduction of instruction cache misses is reported for one benchmark. All other parameters (instruction and data memory accesses, data cache misses) are worse after the transformation. Unfortunately, the results are generated with cache simulation software which is known to be imprecise. The effects of the optimizations on the runtime of the benchmark are not reported.

In the context of optimizations for memory hierarchy exploitation, research has concentrated on the efficient cache utilization for a long period. Code transformations for data flow optimization (e. g. *loop tiling, loop interchange* or *loop fusion*) focus on loop nests since applications spend most of their runtime in the innermost loops. Generally, the iteration space of loop nests is reordered in such a way that a higher locality of data accesses is achieved. The higher the spatial and temporal locality of data accesses, the fewer cache misses occur during program execution resulting in a more efficient cache utilization. Even nowadays, locality optimization is an area of ongoing research. In [Fraboulet et al., 1999], loops are aligned so that the time between two successive accesses to the same memory location (*temporal locality*) is minimized. A graph based optimization strategy is used in [Kandemir, 2002] to cluster array references in loops with spatial or temporal reuse. This technique leads to average reductions of cache misses by 13.8%.

Array padding [Bacon et al., 1994] is a good example of a data layout trans-formation. Here, unused data locations are inserted between columns of an array so as to reduce cache set conflicts and cache thrashing. An approach for simultaneous generation of optimized data layouts and temporal locality improvement is presented in [Loechner et al., 2002]. In this article, geomet-ric models and algorithms are used to minimize TLB (*Translation Look-aside Buffer*) misses.

5.3   Analysis and Optimization Techniques for Loop Nest Splitting

In this section, the analysis and optimization techniques required for loop nest splitting are presented.

An overview of the entire design flow for the analysis of loop nests and the final source code transformation is given in figure 5.6. Since the analysis techniques presented in this chapter require that the source code to be optimized meets some preconditions, these requirements are checked in the very beginning of the design flow. During this phase labeled "Structural C Code Analysis" in figure 5.6, only suitable loop nests and *if*-statements are extracted from the source code. The output of this phase consists of a set of polytopes. Due to the very technical character of this phase, only the preconditions mentioned above are defined formally in section 5.3.1, whereas the algorithmic details are omitted for the sake of readability.

The core optimization algorithm consists of four sequentially executed tasks that are illustrated as a shaded region in figure 5.6. In the beginning, all con-ditions in a loop nest are analyzed separately without considering any inter-dependencies among them. First, it is detected if conditions ever evaluate to true or not (*"condition satisfiability"*, see section 5.3.2). Second, all satisfi-able conditions are analyzed and an optimized search space for each condition is constructed (*"condition optimization"*, cf. section 5.3.3). In a third step, all local search spaces are combined to form a global search space (*"global*

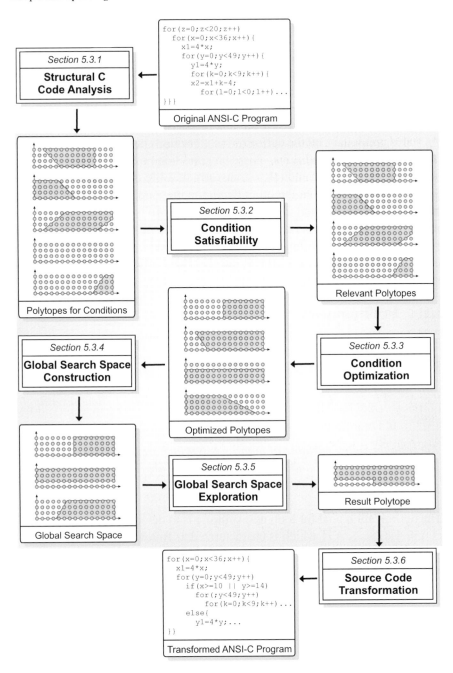

Figure 5.6. Design Flow of Loop Nest Splitting

search space construction", section 5.3.4) which has to be explored leading to
the optimized result for loop nest splitting (*"global search space exploration"*,
compare section 5.3.5).

The result of the search space exploration phase is finally used to rewrite a
loop nest (*"source code transformation"*, see section 5.3.6). For this purpose,
the splitting *if*-statement has to be generated and inserted in the loop nest. Its
then- and *else*-parts are created by replicating parts of the loop nest.

As stated previously, all the optimization algorithms presented in this section
are based on a few preconditions. These preconditions are formulated such that
all algorithms are general enough to transform real-life data dominated source
codes on the one hand. On the other hand, the preconditions help in keeping
some algorithms and formal details simple and easy to understand. In order to
provide the reader with a good understanding of loop nest splitting, extensions
of the optimization algorithms relaxing some of these preconditions are not
directly integrated into this section. Instead, these extensions are described
separately in section 5.4.

5.3.1   Preliminaries

A loop nest is characterized by the following definition.

DEFINITION 5.1 (LOOP NEST)

a *Let* $\Lambda = \{L_1, \ldots, L_N\}$ *be a **loop nest** of depth* N, *where* L_l *denotes a single
   loop.*

b *Let* i_l *be the **index variable** of loop* $L_l \in \Lambda$ *which ranges between its lower
   and upper **bounds*** lb_l *and* ub_l *resp.:* $lb_l \leq i_l \leq ub_l$ ($lb_l, i_l, ub_l \in \mathbb{Z}$). *After
   every iteration of* L_l, i_l *is modified by a **stride*** $s_l \in \mathbb{Z}$.

The optimization goal for loop nest splitting is to determine a set of *alternatives* $\alpha = \{A_1, \ldots, A_J\}$ which is characterized as follows:

DEFINITION 5.2 (OPTIMIZATION CRITERIA)

a *An **alternative*** $A_j \in \alpha$ *is a set of* N *pairs of values*

$$A_j = \{(lb_1', ub_1'), (lb_2', ub_2'), \ldots, (lb_N', ub_N')\}$$

b *For a loop* $L_l \in \Lambda$, *a pair of values* $(lb_l', ub_l') \in A_j$ *specifies a **range of
   values** for the index variable* i_l. *Therefore,* lb_l' *and* ub_l' *must be within the
   loop bounds of* L_l: $lb_l' \geq lb_l$ *and* $ub_l' \leq ub_l$.

c *For all ranges specified by an alternative A_j, all loop-variant if-statements (i. e. if-statements depending on the index variables) in Λ are satisfied:*

$$\forall\, (i_1, \ldots, i_N) \in \mathbb{Z}^N : \quad lb'_1 \leq i_1 \leq ub'_1, \ldots, lb'_N \leq i_N \leq ub'_N \text{ and}$$
$$(lb'_1, ub'_1), \ldots, (lb'_N, ub'_N) \in A_j \Rightarrow$$

all loop-variant *if*-statements are satisfied.

d *Loop nest splitting using all alternatives $A_j \in \alpha$ leads to the minimization of* if-*statement execution.*

All pairs of values (lb'_l, ub'_l) of an alternative $A_j \in \alpha$ specify maximum ranges of values for all index variables where all *if*-statements are true. These ranges and alternatives are used for the construction of the conditions of the splitting *if*-statement as shown in the following example 5.2.

EXAMPLE 5.2

For the code shown in figure 5.1, the techniques presented in this chapter compute the following set of alternatives $\alpha = \{A_1, A_2\}$:

$$A_1 = \{(0, 19), (10, 35), (0, 48), (0, 8), (0, 8), (0, 4), (0, 4)\}$$
$$A_2 = \{(0, 19), (0, 35), (14, 48), (0, 8), (0, 8), (0, 4), (0, 4)\}$$

As can be seen, all pairs (lb'_l, ub'_l) included in α represent the corresponding upper and lower loop bounds of loop L_l except the second pair of A_1 and the third one of A_2. For this reason, only the index variables x and y of the second and third loop of the loop nest are considered for the creation of the splitting-if.

Since α is defined to be a set of alternatives, the conditions generated out of A_1 and A_2 are connected using the logical OR operator leading to the following splitting if-statement which can also be found in figure 5.3:

```
if (x >= 10 || y >= 14)
```

For a hypothetical set of alternatives $\alpha' = \{A_1, A_2\}$

$$A_1 = \{(0, 19), (10, 31), (0, 48), (0, 8), (0, 8), (0, 4), (0, 4)\}$$
$$A_2 = \{(0, 19), (0, 35), (14, 42), (0, 8), (0, 8), (0, 4), (0, 4)\}$$

the following splitting-if would be generated:

```
if ((x >= 10 && x <= 31) || (y >= 14 && y <= 42))
```

Although α' is also a valid solution satisfying criteria a to c of definition 5.2, it is not considered for loop nest splitting since α' does not minimize if-statement executions. As can be seen, the pairs $(10, 31)$ and $(14, 42)$ of α' define sub-intervals of the pairs $(10, 35)$ and $(14, 48)$ of α. This implies that loop nest splitting using α' does not consider the values $32 \leq x \leq 35$ and $43 \leq y \leq 48$ so that a minimization of if-statement execution does not take place for these intervals. Hence, the pairs of α' do not specify maximum ranges of values for the index variables.

The techniques described in the following require that some preconditions are met.

PRECONDITIONS:

1. All loop bounds lb_l and ub_l are constants.

2. All *if*-statements have the format if $(C_1 \oplus C_2 \oplus \dots)$ where C_x are loop-variant conditions that are combined with logical operators $\oplus \in \{\&\&, | \,| \}$. A condition is said to be loop-variant if it only depends on index variables of the loop nest.

3. Loop-variant conditions C are affine expressions[1] of i_l and can have the form $C \simeq \sum_{l=1}^{N}(c_l' * i_l) + c' \otimes \sum_{l=1}^{N}(c_l'' * i_l) + c''$ for constant values c_l', c_l'', c', $c'' \in \mathbb{Z}$ and comparators $\otimes \in \{<, \leq, >, \geq\}$. For the sake of simplicity, only the following normalized equivalent format of conditions is considered from now on, where $C = \sum_{l=1}^{N}(c_l * i_l) \geq c$ for constants $c_l, c \in \mathbb{Z}$.

Precondition 2 is formulated only for keeping the following algorithms and formulas simple and thus does not impose a general restriction on the optimization approach itself. According to the rule of *de Morgan*, the expression $! (C_1$ && $C_2)$ is equivalent to $(\overline{C_1} \,|\,| \; \overline{C_2})$ ($! (C_1 \,|\,| \; C_2)$ analogously). By inverting the comparators in $\overline{C_1}$ and $\overline{C_2}$, the logical *NOT* can also be modeled in *if*-statements. Since all boolean functions can be expressed with &&, | | and !, precondition 2 does not constrain the way how conditions can be combined.

A condition like (a == b) can be rewritten as (a $\geq$ b) && (a $\leq$ b) without loss of generality ((a != b) analogous), so that the set $\otimes$ of supported comparators defined in precondition 3 is not a restriction either.

Forward substitution [Muchnick, 1997] is a well-known code transformation which replaces uses of a variable by its defining expression. By application of this optimization to induction variables not being an index variable i_l of the loop nest Λ (e. g. x1, ..., x4, y1, ..., y4 in figure 5.1), it is assured that C only depends on the index variables i_l as stated in precondition 3 (see also the following example 5.3)[2].

---

[1] F is said to be an affine function if there is a linear function F' and a vector v such that $F(x) = F'(x) + v$. Hence, an affine function is just a linear function plus a translation.

[2] Forward substitution is only applied to internal data structures so as to obtain the required structure of *if*-statements. After the analysis for loop nest splitting is done, the effects of forward substitution are undone so that the source code generated after loop nest splitting exactly looks as illustrated in figure 5.3. This way it is guaranteed that loop nest splitting does not revoke previously applied optimizations (e. g. common subexpression elimination).

EXAMPLE 5.3

After having applied forward substitution to the code shown in figure 5.1, the if-statements have the following form:

```
if (4*x + i < 0 || 35 < 4*x + i ||
    4*y + j < 0 || 48 < 4*y + j)
if (4*x + k + i - 4 < 0 || 35 < 4*x + k + i - 4 ||
    4*y + 1 + j - 4 < 0 || 48 < 4*y + 1 + j - 4)
```

It is obvious that these if-statements *meet precondition 2. For obtaining the normalized format mentioned in precondition 3, the conditions of the if-statements are translated to the following appearance which will serve as example throughout the remainder of this chapter:*

```
if (-4*x - i >= 1 || 4*x + i >= 36 ||
    -4*y - j >= 1 || 4*y + j >= 49)
```

respectively

```
if (-4*x - k - i >= -3 || 4*x + k + i >= 40 ||
    -4*y - 1 - j >= -3 || 4*y + 1 + j >= 53)
```

5.3.2 Condition Satisfiability

In the first phases of the optimization algorithm, all affine conditions C in a loop nest are analyzed separately. Every single condition defines a particular subset of the *total iteration space* of a loop nest Λ which is defined by the following polytope:

DEFINITION 5.3 (TOTAL ITERATION SPACE)
Let Λ be a loop nest of depth N. A matrix B having $2N$ rows and N columns and a vector b of length $2N$ are defined as follows:

$$
B = \begin{pmatrix}
1 & 0 & \cdots & 0 \\
-1 & 0 & \cdots & 0 \\
0 & 1 & \cdots & 0 \\
0 & -1 & \cdots & 0 \\
\vdots & \vdots & \ddots & \vdots \\
0 & 0 & \cdots & 1 \\
0 & 0 & \cdots & -1
\end{pmatrix}, \quad
b = \begin{pmatrix}
lb_1 \\
-ub_1 \\
lb_2 \\
-ub_2 \\
\vdots \\
lb_N \\
-ub_N
\end{pmatrix}
$$

*Using B and b, the polytope T modeling the **total iteration space** is defined by:*

$$
T = \{\, x \in \mathbb{Z}^N \mid Bx \geq b \,\}
$$

For every loop L_l, B and b contain two rows representing constraints on the index variable i_l. The first one ensures that i_l is always greater than or equal to the loop's lower bound, the second constraint represents the upper loop bound. This way, T forms an N-dimensional geometric space limited by all loop bounds lb_l and ub_l and thus models all iterations of the entire loop nest.

Since an affine condition $C = \sum_{l=1}^{N}(c_l * i_l) \geq c$ is nothing but an additional constraint imposed on the total iteration space, a polytope modeling the iterations of the loop nest under consideration of C can be generated easily. For this purpose, the matrix B and vector b introduced in definition 5.3 only need to be extended by one additional row. In the case of the matrix B, this new row contains all the constant factors c_l of condition C, whereas the new element of vector b is set to the constant c. This polyhedral model for loop-variant conditions is illustrated in the following example 5.4.

EXAMPLE 5.4

The condition 4*x + 3*i >= 36, *where the index variables* x *and* i *iterate through the intervals* $[0, 35]$ *and* $[0, 3]$ *resp., has the following polytope representation:*

$$P = \left\{ x \in \mathbb{Z}^2 \middle| \begin{pmatrix} 4 & 3 \\ 1 & 0 \\ -1 & 0 \\ 0 & 1 \\ 0 & -1 \end{pmatrix} x \geq \begin{pmatrix} 36 \\ 0 \\ -35 \\ 0 \\ -3 \end{pmatrix} \right\}$$

The first constraint of P represents the condition itself, whereas the remaining four inequations constrain the polytope to the intervals of x *and* i *mentioned above. Transformed into a graphical illustration, the condition above defines the shape shown in figure 5.7.*

Figure 5.7.   Conditions as Polytopes

According to this scheme, the system of linear inequations for every affine condition C is generated. To these inequations, the Motzkin algorithm and its extensions for the removal of redundant constraints mentioned in section 3.1 is applied for creating the polytope representation P of a condition C.

After that, it can be determined in constant time if the number of equations $Ax = a$ of P (see definition 3.1a on page 25) is equal to the dimension of P plus 1. If this is the case, P is overconstrained and thus defines the empty set as proven in [Wilde, 1993]. As a consequence, C is shown to be unsatisfiable. If instead, P only contains the constraints modeling the loop bounds lb_l and ub_l, P is equal to the total iteration space so that the condition C is shown to be satisfied for all values of the index variables i_l.

Redundant conditions that have been proven to be satisfied or unsatisfied for all iterations of the loop nest are not considered during the next steps of optimization. Instead, the information about the satisfiability of these conditions is reused only in the final phase of the optimization process during source code transformation. Here, all redundant conditions are removed from the source code and are replaced by their according truth value in the *if*-statements (see section 5.3.6).

5.3.3   Condition Optimization

In the next step of the optimization process, all non-redundant conditions $C = \sum_{l=1}^{N}(c_l * i_l) \geq c$ are optimized one after the other. A condition C is analyzed in isolation without considering any other conditions in the loop nest Λ. During this stage, it is assumed that Λ contains only one *if*-statement consisting of exactly one condition, namely C. Without loss of generality, the *if*-statement containing C is located in the innermost loop L_N.

The goal of this stage is to determine a local optimum of *if*-statement executions after loop nest splitting when considering only Λ and C. The subsequent optimization phases (see sections 5.3.4 and 5.3.5) deal with combining these local optima for all conditions and *if*-statements and generating a globally optimized solution.

The outcome of the optimization of a single condition C is a polytope P_C representing a maximal part of the total iteration space where condition C is provably satisfied. This polytope is built using a set of numerical values that are generated by a genetic algorithm which is applied to C. Since the optimization performed by the genetic algorithm is similar to that one of the satisfiability of integer linear constraints under simultaneous optimization of an objective function, it is obvious that the problem of condition optimization is *NP*-complete.

Since polytopes are defined to be systems of linear equations and inequations, it is not obvious that a genetic algorithm is employed to optimize a condition which is also a linear inequation. Instead, the application of techniques from the domain of integer linear programming appears to be more intuitive. But due to the fact that the objective function to be maximized is non-linear, integer linear programming is not practicable. A linearization of the objective function might be possible but probably leads to an explosion of constraints and decision variables implying high runtimes in order to solve the integer linear program. Genetic algorithms, on the other hand, have proven to be effective in optimizing especially non-linear problems (see section 3.2).

In the next subsections, the chromosomal representation, the fitness function and the way how the polytope P_C is generated out of an individual of the genetic algorithm are described.

5.3.3.1   Chromosomal Representation

In order to define how the genetic algorithm for condition optimization interprets the chromosomes of its individuals, the notion of *relevant loops* is required.

DEFINITION 5.4 (RELEVANT LOOP)

Let $\Lambda = \{L_1, \ldots, L_N\}$ *be a loop nest,* $C = \sum_{l=1}^{N} (c_l * i_l) \geq c$ *a condition. The set* $\ell = \{ \, l \mid L_l \in \Lambda \text{ and } c_l \neq 0 \, \}$ *contains the indices of all **relevant loops** for* C.

The set ℓ of relevant loops obviously defines those loops whose index variables C directly depends on. Any modification of an index variable contained in ℓ has a direct influence on C. Since all loop bounds are constant values (see precondition 1), no interdependencies between index variables themselves can exist. As a consequence, the situation where the modification of an index variable i_l with $c_l = 0$ and thus $l \notin \ell$ influences C can not occur. For this reason, the set ℓ completely defines all relevant loops a condition C depends on[3].

In analogy to the pairs of values introduced in definition 5.2, the genetic algorithm presented in this section determines pairs of values $(lb'_{C,l}, ub'_{C,l})$ for every loop L_l with $l \in \ell$. Once again, these pairs of values specify maximum intervals for the index variables of all relevant loops where condition C is satisfied. As a consequence, $lb'_{C,l}$ and $ub'_{C,l}$ have to be within the loop bounds lb_l and ub_l. For all loops $L_l, l \notin \ell$, C does not depend on, the pairs $(lb'_{C,l}, ub'_{C,l})$ are set to the lower and upper loop bounds lb_l and ub_l. All $lb'_{C,l}$ and $ub'_{C,l}$ are determined in such a way that loop nest splitting of Λ according to all pairs would minimize the number of executions of C.

In order to achieve an efficient chromosomal representation of the pairs $(lb'_{C,l}, ub'_{C,l})$ and in order to reduce the search space for the genetic algorithm, the notion of the *monotony of conditions* is important.

THEOREM 5.1 (MONOTONY OF AFFINE CONDITIONS)

Every condition $C = \sum_{l=1}^{N} (c_l * i_l) \geq c$ *is an affine monotone function of the index variables of* Λ.

Proof

A condition $\sum_{l=1}^{N} (c_l * i_l) \geq c$ is a mapping $C : \mathbb{Z}^N \rightarrow \{false, true\}$ depending on the index variables i_l of a loop nest Λ. In the following, it is assumed without

---

[3] For a relaxation of precondition 1 and the implications, refer to section 5.4.

loss of generality that *false* < *true* holds so that an ordering of the truth values is given. The affinity of C is given by construction.

Let $x = (x_1, \dots, x_N) \in \mathbb{Z}^N$ be some concrete assignment of values to the index variables. Consequently, $C(x)$ denotes the well-defined truth value when applying C to x. Without loss of generality, $c_1 \neq 0$ is assumed in the following. Let $x' = (x'_1, x'_2, \dots, x'_N)$ be an alternative assignment of values to the index variables, where only the value for the first index variable i_1 is modified: $x'_1 \neq x_1$ and $x'_2 = x_2, \dots, x'_N = x_N$. Only one of the following four cases can occur:

1. $x'_1 > x_1, c_1 > 0$: The value of $\sum_{l=1}^{N} (c_l * x'_l)$ is greater than $\sum_{l=1}^{N} (c_l * x_l)$.

   If $C(x)$ is *false*, $C(x')$ can still be *false* or can be *true*, depending on the actual values of x'_1 and c.
   If $C(x)$ is *true*, $C(x')$ must also be *true*.
   $\Rightarrow C$ is monotonically increasing in terms of i_1.

2. $x'_1 > x_1, c_1 < 0$: The value of $\sum_{l=1}^{N} (c_l * x'_l)$ is smaller than $\sum_{l=1}^{N} (c_l * x_l)$.

   If $C(x)$ is *true*, $C(x')$ can either be *true* or *false*.
   If $C(x)$ is *false*, $C(x')$ must also be *false*.
   $\Rightarrow C$ is monotonically decreasing in terms of i_1.

3. $x'_1 < x_1, c_1 > 0$: The value of $\sum_{l=1}^{N} (c_l * x'_l)$ is smaller than $\sum_{l=1}^{N} (c_l * x_l)$.

   If $C(x)$ is *true*, $C(x')$ can either be *true* or *false*.
   If $C(x)$ is *false*, $C(x')$ must also be *false*.
   $\Rightarrow C$ is monotonically decreasing in terms of i_1.

4. $x'_1 < x_1, c_1 < 0$: The value of $\sum_{l=1}^{N} (c_l * x'_l)$ is greater than $\sum_{l=1}^{N} (c_l * x_l)$.

   If $C(x)$ is *false*, $C(x')$ can either be *false* or *true*.
   If $C(x)$ is *true*, $C(x')$ must also be *true*.
   $\Rightarrow C$ is monotonically increasing in terms of i_1.

Since all possible cases are covered above and since this argumentation is not limited only to x_1 but is also valid for all values $x_1, \dots, x_N$, a condition C is a monotone function of all index variables i_l.

□

Due to the monotony of a condition C, it is not necessary to deal with pairs of values $(lb'_{C,l}, ub'_{C,l})$. If C is true for a certain value $v \in [lb'_{C,l}, ub'_{C,l}]$ and $c_l > 0$, then C must also be true for $v+1, v+2, \dots, ub_l$. It is impossible that C is true for some v and false for $v+1$. The same holds for $c_l < 0$; in this case, C must be true for $v, v-1, v-2, \dots, lb_l$. Since it is the goal of the genetic algorithm to determine maximum ranges of iterations where a condition C is true, $ub'_{C,l}$ can implicitly be set to the upper loop bound ub_l for range maximization if $c_l > 0$. Analogously, $lb'_{C,l}$ is assumed to be equal to lb_l if $c_l < 0$.

As a consequence of this implicit setting of one value of a pair $(lb'_{C,l}, ub'_{C,l})$ to the corresponding lower or upper loop bound, the following chromosomal representation is used by the genetic algorithm for condition optimization:

DEFINITION 5.5 (CHROMOSOMAL REPRESENTATION)

Let $C = \sum_{l=1}^{N} (c_l * i_l) \geq c$ be a condition and $\ell = \{\, l \mid L_l \in \Lambda$ and $c_l \neq 0 \,\}$ be the set of indices of all relevant loops of C.

a *A **chromosome** is an array of integer values of length $|\ell|$.*

b *For every loop L_l with $l \in \ell$, a **gene** of the chromosome represents a value denoted as $v'_{C,l}$.*

 If $c_l > 0$, $v'_{C,l}$ is interpreted in such a way that condition C should be satisfied for all values of the index variable i_l within the interval $[v'_{C,l}, ub_l]$. If instead $c_l < 0$, condition C is assumed to be satisfied for $i_l \in [lb_l, v'_{C,l}]$.

 The interval $[v'_{C,l}, ub_l]$ or $[lb_l, v'_{C,l}]$ specified by $v'_{C,l}$ is termed $R_{C,l}$.

c *For every gene of a chromosome, the **domain** of possible values for the corresponding variable $v'_{C,l}$ is constrained in such a way that $v'_{C,l}$ must be within the bounds of loop L_l: $lb_l \leq v'_{C,l} \leq ub_l$.*

In contrast to a chromosomal representation where two genes are allocated for a loop L_l with $l \in \ell$ representing the entire pair $(lb'_{C,l}, ub'_{C,l})$, the encoding presented in definition 5.5 is more efficient since the naïve approach results in chromosomes of double length and thus increases the runtimes of the genetic algorithm presented in the following. Since the monotony of a condition C has led to the result that always exactly one bound of the interval $[lb'_{C,l}, ub'_{C,l}]$ is redundant, a naive chromosomal encoding would implicitly contain this kind of redundancy implying an unnecessarily enlarged search space. For this reason, definition 5.5 is not only efficient in terms of the chromosome length, but also with respect to the size of the search space which has to be explored and thus leads to an improved convergence of the genetic algorithm.

Even though definition 5.5c ensures that all values stored in a chromosome are within the domains specified by the lower and upper loop bounds, this is not a guarantee that condition C is satisfied for all values within the intervals $[lb'_{C,l}, ub'_{C,l}]$. Individuals implying that C is not satisfied for the entire interval are called *invalid individuals*:

DEFINITION 5.6 (INVALID INDIVIDUAL)

Let $C = \sum_{l=1}^{N}(c_l * i_l) \geq c$ *be a condition and* $I = (v'_{C,1}, \ldots, v'_{C,|\ell|})$ *an individual of the genetic algorithm.* I *is said to be an **invalid individual** if*

$$C(v'_{C,1}, \ldots, v'_{C,|\ell|}) = false$$

In order to determine whether a given individual I is invalid or not, only the sum $\sum_{l\in\ell}(c_l * v'_{C,l})$ needs to be computed. If this sum is smaller than c, condition C is obviously invalid. If instead the sum is greater than or equal to c, C must be satisfied for all loop nest iterations $R_{C,l}$ specified by I due to the monotony of C.

EXAMPLE 5.5

In this example, the condition $C = $ 4*x + k + i >= 40 *taken from example 5.3 is considered. Since* C *only depends on the index variables* x, k *and* i, *a chromosome length of 3 is required for the optimization of* C. *During the execution of the genetic algorithm, an individual* I *having the chromosome* $(10, 0, 0)$ *can possibly be generated. The numbers encoded in* I *denote the values* $v'_{C,x}$, $v'_{C,k}$ *and* $v'_{C,i}$:

$$v'_{C,x} = 10 \quad v'_{C,k} = 0 \quad v'_{C,i} = 0$$

It is easy to see that C *is satisfied for* x $= 10$, k $= 0$ *and* i $= 0$. *Due to the monotony of* C *and due to the fact that* x *is multiplied by the positive constant 4,* C *must necessarily also be satisfied for all values of* x *greater than 10. In more detail,* C *must be satisfied for all values of* x *between 10 and 35 which is the upper bound of the x-loop. For this reason, the intervals* $R_{C,l}$ *for the entire loop nest shown in figure 5.1 and condition* C *are defined as follows:*

$$z \in [0, 19] \quad x \in [10, 35] \quad y \in [0, 48] \quad k \in [0, 8] \quad 1 \in [0, 8] \quad i \in [0, 3] \quad j \in [0, 3]$$

Since condition C *is not influenced by the index variables* z, y, 1 *and* j, *the intervals* $R_{C,l}$ *of these four index variables are not computed explicitly by the genetic algorithm, but they are set to the corresponding lower and upper loop bounds.*

For a hypothetical condition $C' = $ -4*x + k + i >= -40, *the individual* $I = (10, 0, 0)$ *results in* $R_{C',x} = [0, 10]$. *This is due to the fact that* C' *is satisfied for* x $= 10$ *and for all values of* x *smaller than 10 because of the monotony and the negative factor of* -4.

5.3.3.2   Fitness Function

For the optimization of an affine condition C using a genetic algorithm, a *fitness function* is required to measure the quality of an individual I. Given an individual I which is represented by a chromosome according to definition 5.5, the fitness of I is the higher, the fewer *if*-statements would be executed when splitting loop nest Λ according to the values $v'_{C,l}$ encoded in I. In the case that an invalid individual is generated during genetic optimization, a very low fitness value is assigned to this individual. Since the genetic algorithm only selects those individuals in a population having the highest fitness values, this genetic optimization consequently leads to a minimization of the number of executed *if*-statement executions.

Given a condition $C = \sum_{l=1}^{N}(c_l * i_l) \geq c$ and a set of values $v'_{C,l}$ stored in a valid individual of the genetic algorithm, it is necessary to compute the number of executed *if*-statements after loop nest splitting for the evaluation of the fitness of an individual I. For these computations, the following values are required.

DEFINITION 5.7

Let $\Lambda = \{L_1, \ldots, L_N\}$ *be a loop nest,* $C = \sum_{l=1}^{N}(c_l * i_l) \geq c$ *a condition and* I *a valid individual.*

a *The **size of the iteration space of** Λ IS_Λ represents the total number of executions of the body of loop L_N under consideration of surrounding loops and loop strides:*

$$IS_\Lambda = \prod_{l=1}^{N}(\lfloor(ub_l - lb_l) / \text{abs}(s_l)\rfloor + 1)$$

b *The **iteration space** of individual I represents the total number of executions of the body of loop L_N under consideration of the iteration ranges defined by I. The size of the iteration space of I IS_I is equal to the iteration space of Λ reduced to the intervals represented by $v'_{C,l}$:*

$$IS_I = \prod_{l=1}^{N} r_l \text{ and } r_l = \begin{cases} \lfloor(ub_l - lb_l) / \text{abs}(s_l)\rfloor + 1 & \text{if } c_l = 0, \\ \lfloor(ub_l - v'_{C,l}) / \text{abs}(s_l)\rfloor + 1 & \text{if } c_l > 0, \\ \lfloor(v'_{C,l} - lb_l) / \text{abs}(s_l)\rfloor + 1 & \text{otherwise} \end{cases}$$

c *The **innermost loop** index λ refers to the maximum loop for which the iteration ranges defined by $v'_{C,\lambda}$ are true sub-intervals of the loop bounds:*
$$\lambda = \max\{ l \mid L_l \in \Lambda, \ R_{C,l} \subset [lb_l, \ ub_l] \}$$

d *The genetic algorithm assumes that Λ would be **split** in the following way:*

```
for (i₁=lb₁; i₁<=ub₁; i₁+=s₁)
    ...
        for (iλ₋₁=lbλ₋₁; iλ₋₁<=ubλ₋₁; iλ₋₁+=sλ₋₁)
            for (iλ=lbλ; iλ<=ubλ; iλ+=sλ)
                if (i₁ ∈ R_C,1 && ... && iλ ∈ R_C,λ)
                    for (; iλ<=new_boundλ; iλ+=sλ)
                        for (iλ₊₁=lbλ₊₁; iλ₊₁<=ubλ₊₁; iλ₊₁+=sλ₊₁)
                            ...
```

```
                for (iN=lbN; iN<=ubN; iN+=sN)
          else
            for (iλ+1=lbλ+1; iλ+1<=ubλ+1; iλ+1+=sλ+1)
            ...
                for (iN=lbN; iN<=ubN; iN+=sN)
                if (C) { ... };
```

The value of new_bound$_\lambda$ is equal to ub$_\lambda$ for $c_\lambda > 0$ resp. $v'_{C,\lambda}$ for $c_\lambda < 0$.

EXAMPLE 5.6

To continue example 5.5, the condition $C = $ 4*x + k + i $\;>=\;$ 40 and the individual $I =$ (10, 0, 0) are treated again resulting in the intervals $R_{C,l}$

$z \in [0, 19]$ $x \in [10, 35]$ $y \in [0, 48]$ $k \in [0, 8]$ $1 \in [0, 8]$ $i \in [0, 3]$ $j \in [0, 3]$

For the entire loop nest depicted in figure 5.1, the body of the j loop is executed 45,722,880 times since $IS_\Lambda = 20 * 36 * 49 * 9 * 9 * 4 * 4 = 45,722,880$. If x is constrained to the interval [10, 35], the j loop is executed 33,022,080 times in total: $IS_I = 20 * 26 * 49 * 9 * 9 * 4 * 4 = 33,022,080$.

The fitness function for condition optimization supposes that loop nest splitting according to I would be performed leading to the following hypothetical code fragment:

```
for (z=0; z<20; z++)
  for (x=0; x<36; x++)
    if (x >= 10)                         /* Splitting-If */
      for (; x<36; x++)                  /* Duplicated Loop Lλ */
        for (y=0; y<49; y++)
        ...
            for (j=0; j<4; j++) { ... }
    else
      for (y=0; y<49; y++)
      ...
          for (j=0; j<4; j++)
            if (4*x + k + i >= 40) ...;   /* Original If */
```

Since only the iteration space of the index variable x is constrained by I, the innermost loop λ refers to the x loop. λ can therefore be interpreted as the index of the loop containing the splitting-if defined by I.

For a hypothetical code after loop nest splitting using the iteration ranges $R_{C,l}$ defined by an individual I, the fitness function has to compute how many times *if*-statements are executed in total. To compute this value, denoted as IF_{Tot}, it has to be taken into account that the loop nest Λ is assumed to contain two different kinds of *if*-statements (see the code from example 5.6): First, the number of executions of the original *if*-statement "if (C)" in the innermost loop L_N have to be counted (termed IF_{Orig}). Second, the number of times the splitting-if is executed (called IF_{Split}) has to be added.

COROLLARY 5.1 $IF_{Total} = IF_{Orig} + IF_{Split}$

The original *if*-statement is only executed if the splitting-if evaluates to false. Since all loop nest iterations with a satisfied splitting-if are denoted as IS_I, this value has to be subtracted from IS_Λ to obtain IF_{Orig}.

COROLLARY 5.2 $IF_{Orig} = IS_\Lambda - IS_I$

To compute the number of executions of the splitting-if IF_{Split}, more effort has to be spent due to the duplication of the innermost loop L_λ (e. g. the x loop in example 5.6). As a starting point, the observation is used that an *if*-statement is executed as often as its *then*- and *else*-parts are executed in total. The number of executions of *then*- and *else*-part of the splitting *if*-statement are denoted TP_{Split} resp. EP_{Split}.

COROLLARY 5.3 $IF_{Split} = TP_{Split} + EP_{Split}$

IS_I represents the number of times the body of loop L_N is executed when the splitting-if is true. After dividing IS_I by the number of iterations of loop L_N, the resulting quotient represents the number of times the body of loop L_{N-1} is executed for a satisfied splitting-if. Analogously, the additional division by the number of iterations of loop L_{N-1} leads to the execution frequency of loop L_{N-2}. By repeating these divisions for all loops $L_{\lambda+1}, \ldots, L_N$, the number of executions of loop L_λ for a satisfied splitting-if is computed. Still, the duplication of loop L_λ in the *then*-part of the splitting-if is not yet considered. For this purpose, the division by the value r_λ (see definition 5.7b) is sufficient, since this value represents the number of iterations of loop L_λ in the case of a satisfied splitting-if.

COROLLARY 5.4 $TP_{Split} = IS_I \left/ \left(\prod\limits_{l=\lambda+1}^{N} (\lfloor (ub_l - lb_l) / \mathrm{abs}(s_l) \rfloor + 1) * r_\lambda \right) \right.$

Using a similar kind of argumentation, EP_{Split} can be determined. In this case, all loop nest iterations IF_{Orig} with a false splitting-if are used. This value needs to be divided by all iterations of the loops $L_{\lambda+1}, \ldots, L_N$ to obtain the number of executions of the *else*-part of the splitting-if.

COROLLARY 5.5 $EP_{Split} = IF_{Orig} \left/ \prod\limits_{l=\lambda+1}^{N} (\lfloor (ub_l - lb_l) / \mathrm{abs}(s_l) \rfloor + 1) \right.$

EXAMPLE 5.7

For the piece of C code depicted in example 5.6, the iteration spaces of Λ and the individual $I = (10, 0, 0)$ amount to 45,722,880 and 33,022,080 resp.

Corollary 5.2 implies that the original if-*statement* if (4*x + k + i >= 40) *located in the* j *loop of the* else-*part of the splitting-if is executed 12,700,800 times, since*

$$IF_{Orig} \quad = \quad 45,722,880 - 33,022,080 \ = \ 12,700,800$$

Corollaries 5.3 – 5.5 lead to the result that the splitting-if if (x >= 10) *is executed only 220 times which illustrates the large benefits of loop nest splitting.*

$$
\begin{aligned}
TP_{Split} &= 33,022,080 \ / \ (49 * 9 * 9 * 4 * 4 * 26) \ = \ 20 \\
EP_{Split} &= 12,700,800 \ / \ (49 * 9 * 9 * 4 * 4) \ = \ 200 \\
IF_{Split} &= 20 + 200 \ = \ 220
\end{aligned}
$$

Finally, the fitness function of the genetic algorithm for condition optimization computes a total of 12,701,020 if-statement executions for the individual I and the code fragment shown above. This result is due to corollary 5.1:

$$IF_{Total} \quad = \quad 12,700,800 + 220 \ = \ 12,701,020$$

When combining the formulas given in the corollaries above, a compact aggregate expression describing the total number of *if*-statement executions after loop nest splitting for an individual I and a condition C can be derived.

THEOREM 5.2 (IF-STATEMENT EXECUTIONS)

Let $\Lambda = \{L_1, \ldots, L_N\}$ *be a loop nest,* $C = \sum_{l=1}^{N}(c_l * i_l) \geq c$ *a condition and I a valid individual. The number of* if-*statement executions according to definition 5.7.d is given by*

$$IF_{Total} \quad = \quad IS_\Lambda - IS_I + \prod_{l=1}^{\lambda-1} r_l + \frac{IS_\Lambda - IS_I}{\displaystyle\prod_{l=\lambda+1}^{N} r_l}$$

Proof

The expression $IS_\Lambda - IS_I$ is equal to IF_{Orig} (see corollary 5.2).

Due to the fact that λ is defined to be the maximum loop index for which the iteration ranges $R_{C,l}$ are true sub-intervals of the loop bounds, it can be concluded that $R_{C,l} = [lb_l, \ ub_l]$ for $l \in [\lambda + 1, \ldots N]$. This implies that r_l must be equal to $\lfloor (ub_l - lb_l) \ / \ abs(s_l) \rfloor + 1$ for $l \in [\lambda + 1, \ldots, N]$. Using these implications and corollary 5.4, TP_{Split} is equal to

$$TP_{Split} \quad = \quad IS_I \ / \ \left(\prod_{l=\lambda+1}^{N} (\lfloor (ub_l - lb_l) \ / \ abs(s_l) \rfloor + 1) * r_\lambda \right)$$

$$= \quad IS_I \Big/ \prod_{l=\lambda+1}^{N} r_l * r_\lambda$$

$$= \quad IS_I \Big/ \prod_{l=\lambda}^{N} r_l$$

$$= \quad \prod_{l=1}^{N} r_l \Big/ \prod_{l=\lambda}^{N} r_l$$

$$= \quad \prod_{l=1}^{\lambda-1} r_l * \prod_{l=\lambda}^{N} r_l \Big/ \prod_{l=\lambda}^{N} r_l$$

$$= \quad \prod_{l=1}^{\lambda-1} r_l$$

In an analog way, EP_{Split} can be expressed by

$$EP_{Split} \quad = \quad \frac{IS_\Lambda - IS_I}{\prod\limits_{l=\lambda+1}^{N} r_l}$$

By inserting the equations above in corollaries 5.3 and 5.1, the formula presented in theorem 5.2 can be derived.

$\square$

Using the formula given in theorem 5.2, the fitness function of the genetic algorithm for condition optimization determines the quality of a valid individual I. As a final step, it has to be ensured that the fitness of a valid individual is the higher, the fewer *if*-statements are executed. Furthermore, the fitness of invalid individuals has to be defined.

DEFINITION 5.8 (FITNESS OF INDIVIDUALS)

Let $\Lambda = \{L_1, \ldots, L_N\}$ *be a loop nest and* $C = \sum\limits_{l=1}^{N} (c_l * i_l) \geq c$ *a condition.*

a *The* ***fitness** of a **valid** individual* I *is*
$(1 \,/\, IF_{Total}) * IS_\Lambda = IS_\Lambda \,/\, IF_{Total}$

b *The* ***fitness** of an **invalid** individual* I *is*
$$\left(1 \,/\, \left(c - \sum_{l=1}^{N} (c_l * v'_{C,l}) + 2 * IS_\Lambda\right)\right) * IS_\Lambda = IS_\Lambda \,/\, \left(c - \sum_{l=1}^{N} (c_l * v'_{C,l}) + 2 * IS_\Lambda\right)$$

By using the inverse $1 \,/\, IF_{Total}$ for the computation of the fitness of a valid individual, the maximization of the fitness by the genetic algorithm implicitly leads to the desired minimization of *if*-statement executions. In order to avoid the generation of too small values by the fitness function, this inverse value is finally multiplied by the large value IS_Λ which is constant during every run of the genetic algorithm.

For an invalid individual I, the condition $C = \sum_{l=1}^{N} c_l * v'_{C,l} \geq c$ is not satisfied implying that $\sum_{l=1}^{N} c_l * v'_{C,l}$ is smaller than c. As a consequence, the expression $c - \sum_{l=1}^{N} c_l * v'_{C,l}$ given in definition 5.8b is a measure for the degree of I's invalidity. This expression serves to distinguish between different invalid solutions so that the genetic algorithm is guided towards a less invalid or even a valid individual. By adding $2 * IS_\Lambda$, it is ensured that each invalid individual always has a lower fitness than any valid solution, since loop nest splitting for a valid individual can never lead to the execution of more than $2 * IS_\Lambda$ *if*-statements. Again, the value IS_Λ is divided by this term so as to obtain the final fitness value.

Due to the fact that a chromosome has a maximum length of N (denoting the depth of the loop nest), and since all elements of the formula for the computation of IF_{Total} are products of N elements, the entire fitness function has linear complexity $\mathcal{O}(N)$.

5.3.3.3 Polytope Generation

After the optimization of a condition C using the genetic algorithm presented above, the individual I_{Opt} having a maximum fitness is used for the generation of a polytope P_C. The set of all polytopes P_C for all conditions C in a loop nest is the final result of the step of condition optimization as illustrated in figure 5.6 on page 63. As described in definition 5.5, every gene of I_{Opt} specifies an interval $R_{C,l}$ for every relevant loop $l \in \ell$. The polytope P_C has to model these intervals under consideration of the loop bounds of the entire loop nest.

DEFINITION 5.9 (OPTIMIZED POLYTOPE)

Let $\Lambda = \{L_1, \ldots, L_N\}$ be a loop nest, $C = \sum_{l=1}^{N}(c_l * i_l) \geq c$ be a condition. $v'_{C,l}$ denotes the values stored in the genes of the individual I_{Opt} with maximum fitness after execution of the genetic algorithm for condition optimization.

*The **polytope P_C** modeling the optimized iteration ranges specified by I_{Opt} is defined as follows:*

$$P_C = \{ (x_1, \ldots, x_N) \in \mathbb{Z}^N \mid \forall L_l \in \Lambda : lb_l \leq x_l \leq ub_l,$$
$$x_l \geq v'_{C,l} \text{ if } c_l > 0,$$
$$x_l \leq v'_{C,l} \text{ if } c_l < 0 \}$$

In analogy to the polytope defining the total iteration space, two constraints are added to P_C for every loop $L_l \in \Lambda$ ensuring that the index variable i_l only ranges between the lower and upper loop bounds lb_l resp. ub_l. Additionally, exactly one constraint is added for every relevant loop of C. If an index variable i_l is multiplied by a positive constant c_l in condition C, the constraint $i_l \geq v'_{C,l}$ is added to P_C. If instead c_l is negative, $i_l \leq v'_{C,l}$ is used. This way, the iteration ranges defined by I_{Opt} are translated exactly into their polyhedral representation.

EXAMPLE 5.8

Assuming that the individual $I = (10, 0, 0)$ *represents the best solution found during the optimization of the condition* $C = 4*x + k + i \geq 40$, *the following intervals* $R_{C,l}$ *already given in example 5.6 have to be modeled by a polytope representation:*

$$z \in [0, 19] \quad x \in [10, 35] \quad y \in [0, 48] \quad k \in [0, 8] \quad 1 \in [0, 8] \quad i \in [0, 3] \quad j \in [0, 3]$$

Out of these particular intervals, the following polytope P_C *can be generated:*

$$P_C = \left\{ (z, x, y, k, 1, i, j) \in \mathbb{Z}^7 \;\middle|\; \begin{array}{lll} 0 \leq z \leq 19, & 10 \leq x \leq 35, & \\ 0 \leq y \leq 48, & 0 \leq k \leq 8, & \\ 0 \leq 1 \leq 8, & 0 \leq i \leq 3, & 0 \leq j \leq 3 \end{array} \right\}$$

5.3.4 Global Search Space Construction

As defined in section 5.3.1, the goal of loop nest splitting is to determine ranges of values of the index variables for which all *if*-statements in a loop nest are satisfied. Since only single conditions of *if*-statements have been treated by the techniques presented so far, the step of global search space construction is the first one dealing explicitly with all conditions and all *if*-statements in a loop nest.

The purpose of this stage is to combine all the polytopes P_C generated by the genetic algorithm presented in the previous section 5.3.3 in such a way that a single finite union of polytopes G results. This finite union of polytopes is called *global search space* and has the following properties:

- G models the polytopes P_C so that the optimized iteration ranges satisfying a single condition C are included in G.

- The structure of G is based on the way how conditions are combined within all *if*-statements of a loop nest.

- The global search space G represents iteration ranges for which all *if*-statements in a loop nest are satisfied.

The generation of the global search space G is performed in a two-stage approach. First, a finite union of polytopes is generated for every single *if*-statement occurring in the loop nest. Second, the finite unions of polytopes for these *if*-statements are combined to model all *if*-statements in the loop nest.

According to precondition 2 formulated in section 5.3.1, an *if*-statement $IF_i = (C_{i,1} \oplus C_{i,2} \oplus \ldots \oplus C_{i,n})$ is a sequence of affine conditions $C_{i,j}$ which are combined with logical operators $\oplus \in \{\&\&, ||\}$. Since the conditions $C_{i,j}$ have been optimized, a polytope $P_{i,j}$ is associated with every condition. In order to build a finite union of polytopes P_i for *if*-statement IF_i, the conditions of IF_i are traversed in their natural execution order π which is defined by the precedence rules for the operators $\&\&$ and $||$. The generation of P_i is defined as follows:

DEFINITION 5.10 (UNION OF POLYTOPES FOR IF-STATEMENTS)
Let $IF_i = (C_{i,1} \oplus C_{i,2} \oplus \ldots \oplus C_{i,n})$ be an if-statement, $P_{i,j}$ be the optimized polytope associated with condition $C_{i,j}$ and π be the permutation of $\{1, \ldots, n\}$ representing the natural execution order of the conditions $C_{i,j}$.

*a   The **finite union of polytopes** P_i is initialized with $P_{i,\pi(1)}$.*

b $\forall j \in \{2, \ldots, n\} : P_i = P_i \uplus P_{i,\pi(j)}$ with

$$\uplus = \begin{cases} \cap & \text{if } C_{i,\pi(j-1)} \,\&\&\, C_{i,\pi(j)} \\ \cup & \text{if } C_{i,\pi(j-1)} \,||\, C_{i,\pi(j)} \end{cases}$$

As can be seen from definition 5.10, the polytopes $P_{i,\pi(j)}$ are successively combined with P_i. Depending on whether the logical *AND* or *OR* operator is used in IF_i for combining $C_{i,\pi(j)}$ with the previous condition $C_{i,\pi(j-1)}$, the intersection or the union operator is applied during this iterative approach. A finite union of polytopes P_i consequently models those ranges of values of the index variables for which a single *if*-statement IF_i is satisfied.

The method described in definition 5.10 is applied to every *if*-statement. Since loop nest splitting deals with the problem of determining ranges of iterations for which all *if*-statements are satisfied, the global search space G is constructed by intersecting all P_i:

DEFINITION 5.11 (GLOBAL SEARCH SPACE)
Let Λ be a loop nest containing k if-statements $IF_1, \ldots, IF_k$. $P_1, \ldots, P_k$ denote the finite unions of polytopes associated with each if-statement.

*The **global search space** G for loop nest splitting is characterized by the following finite union of polytopes:*

$$G = P_1 \cap \ldots \cap P_k$$

EXAMPLE 5.9

In this example, the two if-statements already given in example 5.3 are considered:

$$IF_1 \quad = \quad \begin{aligned} &\texttt{-4*x - i >= 1 || 4*x + i >= 36 ||} \\ &\texttt{-4*y - j >= 1 || 4*y + j >= 49} \end{aligned}$$

$$IF_2 \quad = \quad \begin{aligned} &\texttt{-4*x - k - i >= -3 || 4*x + k + i >= 40 ||} \\ &\texttt{-4*y - l - j >= -3 || 4*y + l + j >= 53} \end{aligned}$$

After the first two steps of the optimization algorithm for loop nest splitting, the following poly-topes are associated with each condition (throughout this example, the constraints modeling all lower and upper loop bounds are omitted for the sake of simplicity):

$$P_{1,1} = \emptyset, \; P_{1,2} = \{x \geq 9\}, \; P_{1,3} = \emptyset, \; P_{1,4} = \{y \geq 13\}$$
$$P_{2,1} = \{x = 0 \wedge k = 0\}, \; P_{2,2} = \{x \geq 10\},$$
$$P_{2,3} = \{y = 0 \wedge l = 0\}, \; P_{2,4} = \{y \geq 14\}$$

According to definition 5.10, the following finite unions of polytopes are created for these two if-statements:

$$P_1 \quad = \quad \{x \geq 9\} \cup \{y \geq 13\}$$
$$P_2 \quad = \quad \{x = 0 \wedge k = 0\} \cup \{x \geq 10\} \cup \{y = 0 \wedge l = 0\} \cup \{y \geq 14\}$$

Since the use of operations on finite unions of polyhedra automatically involves the application of the extended Motzkin algorithm for constraint simplification (see chapter 3.1), the global search space G has the following appearance after the intersection of P_1 and P_2:

$$G \quad = \quad \{x = 0 \wedge k = 0 \wedge y \geq 13\} \cup \{x \geq 10\} \cup$$
$$\{y = 0 \wedge l = 0 \wedge x \geq 9\} \cup \{y \geq 14\}$$

Due to the fact that the illustration of the functionality of Motzkin's algorithm is beyond the scope of this book, only its outcome is shown here. As can be seen, G still is a union of polytopes, each of them based on constraints found during condition optimization.

5.3.5   Global Search Space Exploration

Since the global search space G defined in the previous section is a finite union of polytopes, it can be represented as follows:

$$G = R_1 \cup \ldots \cup R_M$$

Each polytope R_r of G defines a region of the total iteration space where all *if*-statements in a loop nest Λ are satisfied. As a consequence, G already is a valid solution for loop nest splitting, since all *if*-statements are satisfied. But in general, the use of G in its entirety for splitting Λ is not advantageous, because this is not a solution totally minimizing the *if*-statement executions.

EXAMPLE 5.10

When considering the global search space

$$G = \{x = 0 \wedge k = 0 \wedge y \geq 13\} \cup \{x \geq 10\} \cup$$
$$\{y = 0 \wedge l = 0 \wedge x \geq 9\} \cup \{y \geq 14\}$$

presented in the previous example as a whole, the following splitting if-*statement would be generated (compare section 5.3.6):*

```
if ((x == 0 && k == 0 && y >= 13) || (x >= 10) ||
    (y == 0 && l == 0 && x >= 9) || (y >= 14))
```

Since this if-statement depends on the index variable l, *loop nest splitting would have to be performed in the* l-*loop which is the fifth loop in the loop nest depicted in figure 5.1. In total, this way of loop nest splitting would lead to the execution of 10,103,760 if-statements.*

On the other hand, loop nest splitting only using the region $\{x \geq 10\}$ *of G implies the following simple splitting-if:*

```
if (x >= 10)
```

In this situation, a total of 25,401,820 if-statements is executed after loop nest splitting. In contrast, the following splitting-if already shown in figure 5.3

```
if (x >= 10 || y >= 14)
```

leads to the execution of only 7,261,120 if-statements.

For this reason, the final stage of the loop nest split optimization (compare figure 5.6 on page 63) consists of the the selection of appropriate regions R_r of G such that once again the total number of executed *if*-statements is minimized after loop nest splitting. This process of selecting parts of G is called *global search space exploration*. As can be seen from example 5.10, the problem of global search space exploration is highly non-linear due to the disjunction of constraints. For this reason, this task is also performed using genetic algorithms. In the following subsections, the genetic encoding and the used fitness function are described.

5.3.5.1 Chromosomal Representation

Since finite unions of polytopes (i. e. the logical *OR* of constraints) can not be modeled efficiently using integer linear programming, a second genetic algorithm is used for the exploration of G and the selection of appropriate regions R_r. For a given search space $G = R_1 \cup \ldots \cup R_M$, each individual I of a population consists of a bit-vector of length M, where each bit I_r determines whether region R_r of G is selected by the genetic algorithm or not:

DEFINITION 5.12 (CHROMOSOMAL REPRESENTATION)
Let $G = R_1 \cup \ldots \cup R_M$ *be a global search space.*

*a A **chromosome** is an array of integer values of length M.*

b *The **domain** of possible values for every **gene** of a chromosome is limited to 0 and 1.*

A gene I_r equal to 1 represents a region R_r which is selected by the genetic algorithm. For $I_r = 0$, R_r is not selected.

Due to the fact that every polytope R_r of G represents regions of the iteration space for which all *if*-statements in a loop nest are satisfied by construction, a chromosome as specified in definition 5.12 can never represent an invalid individual. Every subset of polytopes R_r of G automatically implies the satisfaction of all *if*-statements. Even the individual $I = (1, \ldots, 1)$ selecting all regions of G is a valid individual, because the entire global search space is a valid solution. The empty set of selected regions is also a valid solution for loop nest splitting, because no splitting-if would be inserted in Λ in this case. Consequently, the situation of invalidity, where the satisfaction of the splitting-if implies that one or more original *if*-statement would not be satisfied, can not occur. Hence, the fitness function of this genetic algorithm does not have to deal with invalid individuals.

DEFINITION 5.13 (SELECTED SEARCH SPACE)
*For an individual I generated by the genetic algorithm, G_I denotes the **selected search space** representing the global search space G reduced to only those regions selected by I:*
$G_I = \bigcup R_r$ with $I_r = 1$

5.3.5.2   Fitness Function

For a given individual $I = (I_1, \ldots, I_M)$, the fitness function of the genetic algorithm for global search space exploration has to compute the number IF_I of *if*-statement executions after loop nest splitting according to G_I. In contrast to the fitness function for condition optimization, no well-defined set of formulas can be given here in order to compute the *if*-statement executions during search space exploration. This is due to the fact that IF_I is closely related to the size of the selected search space G_I. Since it has been shown that the size computation of a finite union of polyhedra is #P-complete [Kaibel and Pfetsch, 2003] in general, no efficient algorithms currently exist. As a consequence, the complexity of the fitness function described in the following is not polynomial, as opposed to the computations performed for condition optimization.

In analogy to section 5.3.3, the index λ of the innermost loop where the loop nest would be split has to be defined for each individual I of the genetic algorithm. In order to be able to count how many *if*-statements are executed when splitting a loop nest using the regions selected by I, some more definitions are necessary:

DEFINITION 5.14 (INNERMOST LOOPS)
Let $G = R_1 \cup \ldots \cup R_M$ be the global search space, $I = (I_1, \ldots, I_M)$ be an individual.

a *Let $R_r = \{ x \mid Ax = a, \ Bx \geq b \}$ be a polytope of G. $A_{i,j}$ denotes the element of matrix A in line i and row j, a_i the i-th element of a ($B_{i,j}$ and b_i analogously).*

*R_r is said to **reference a loop** $L_l \in \Lambda$, if*

- *R_r contains a constraint c with $c = \left(\sum_{i=1}^{N} (A_{i,j} * x_i) = a_i \right)$ and $A_{l,j} \neq 0$*

$\left(c = \left(\sum_{i=1}^{N} (B_{i,j} * x_i) \geq b_i \right) \text{ and } B_{l,j} \neq 0 \text{ resp.} \right)$, and

- *this constraint c not only defines the lower or upper bound of loop L_l, i. e. $c \neq (x_l \geq lb_l)$ and $c \neq (x_l \leq ub_l)$.*

b *λ_r denotes the index of the **innermost loop referenced** by polytope R_r:*
$\lambda_r = \max\{ l \mid L_l \in \Lambda, \ R_r \text{ references } L_l \}$

c *The **sequence** $\mathcal{L} = (\vec{\lambda}_1, \ldots, \vec{\lambda}_M)$ contains all the indices λ_r in ascending order.*

d *The **innermost loop** λ for individual I is the index of the loop where the loop nest has to be split when considering all regions selected by I:*
$\lambda = \max\{ \lambda_r \mid r \in [1, M], \ I_r = 1\}$

EXAMPLE 5.11

In the previous example 5.10, the following global search space has been used (all constraints modeling lower and upper loop bounds are omitted for the sake of simplicity):

$$G = R_1 \cup R_2 \cup R_3 \cup R_4 = \{x = 0 \wedge k = 0 \wedge y \geq 13\} \cup \{x \geq 10\} \cup \{y = 0 \wedge l = 0 \wedge x \geq 9\} \cup \{y \geq 14\}$$

It is easy to see that R_3 references the x-loop, because it contains the constraint $x \geq 9$. Also, R_3 references the l-loop, since $l = 0$ is not a constraint modeling the lower bound of loop 1. But obviously, R_3 does not reference the k-loop. As a consequence, the innermost loops λ_r for the regions $R_1, \ldots R_4$ point to the following loops:

$$\lambda_1 = 4 \, (\hat{=} \, k) \qquad \lambda_2 = 2 \, (\hat{=} \, x) \qquad \lambda_3 = 5 \, (\hat{=} \, l) \qquad \lambda_4 = 3 \, (\hat{=} \, y)$$

As a consequence, the ordered sequence of all loop indices λ_r is given by $\mathcal{L} = (2, 3, 4, 5)$. Assuming that the individual $I = (1, 0, 1, 0)$ is generated by the genetic algorithm, the innermost loop λ refers to the l-loop.

In order to count the number of *if*-statement executions during global search space exploration, the fitness function of the genetic algorithm models the relevant parts of the loop nest after splitting according to an individual I. For this purpose, it is important to compute how many times *if*-statements are executed at a certain position of the loop nest. For a loop L_l, IF_l represents the number of *if*-statement executions of the inner loop nest $L_l, \ldots, L_N$ before splitting:

DEFINITION 5.15 (IF-STATEMENTS IN INNER LOOP NESTS)
Let $\Lambda = \{L_1, \ldots, L_N\}$ be a loop nest.

a *$\#IF_l$ denotes the **number of** **if-statements** which are located in the body of* **loop** *L_l itself but not in any other loop L_l' which is nested within L_l.*

b *IF_l denotes the **number of** **if-statements** that are evaluated when the* **loop** **nest** *$\Lambda' = \{L_l, \ldots, L_N\}$ is executed:*

$$IF_l = (\lfloor (ub_l - lb_l) / \text{abs}(s_l) \rfloor + 1) * (IF_{l+1} + \#IF_l)$$
$$IF_{N+1} = 0$$

EXAMPLE 5.12

For the loop nest depicted in figure 5.1 (see page 54), $\#IF_j$ is equal to 2, all other values $\#IF_l$ are zero. Due to $\#IF_j$ and the loop bounds, the number of if-statement executions within the k-loop $\#IF_k$ is equal to 2,592.

For a global search space $G = R_1 \cup \ldots \cup R_M$, the fitness function is a nest of M loops. Independently of a concrete individual I, only the loops L_{λ_r} are possible candidates where a splitting could be performed. This is due to the fact that a splitting-if for a polytope R_r needs to check the values of the index variables of all loops referenced by R_r. This can only be achieved when placing the splitting-if in the innermost loop L_{λ_r}. For this reason, every loop of the fitness function represents such an innermost loop L_{λ_r}. The loops of the fitness function are nested according to the ordering of the sequence $\mathcal{L}$, so that the outermost loop represents $L_{\vec{\lambda}_1}$, whereas the innermost loop stands for $L_{\vec{\lambda}_M}$. The general structure of the fitness function is depicted as pseudo-code in figure 5.8.

The main goal of this structure of the fitness function is to update a counter `if_count` properly so that it always represents the number of executed *if*-statements IF_I after loop nest splitting under consideration of an individual I. As mentioned in the previous paragraph, every loop of the fitness function (lines 3, 9 and 16 of figure 5.8) represents a relevant loop of Λ where a splitting might be done depending on I. These loops iterate from the lower bound $lb_{\vec{\lambda}_r}$ to the upper bound $ub_{\vec{\lambda}_r}$ using the corresponding loop strides.

```
1    double fitness(individual I) {
2       int iv₁, ..., ivₘ, if_count = 0;

3       for (iv₁=lb_{x̄₁}; iv₁<=ub_{x̄₁}; iv₁+=s_{x̄₁})
4         if (λ == λ̄₁) {
5           if_count++;
6           if (Gᵢ is not satisfied for (iv₁))
7             if_count += IF_{λ+1};
8         } else

9           for (iv₂=lb_{x̄₂}; iv₂<=ub_{x̄₂}; iv₂+=s_{x̄₂})
10            if (λ == λ̄₂) {
11              if_count++;
12              if (Gᵢ is not satisfied for (iv₁, iv₂))
13                if_count += IF_{λ+1};
14            } else

15              ⋮

16              for (ivₘ=lb_{x̄ₘ}; ivₘ<=ub_{x̄ₘ}; ivₘ+=s_{x̄ₘ})
17                if (λ == λ̄ₘ) {
18                  if_count++;
19                  if (Gᵢ is not satisfied for (iv₁, ..., ivₘ))
20                    if_count += IF_{λ+1};
21                } else

22                  if_count += IF_{λ+1};
23    return ((double) (IF₁ / if_count)); }
```

Figure 5.8. Structure of the Fitness Function for Global Search Space Exploration

The first thing to do during every iteration of a loop $L_{\vec{x}_r}$ is to check whether the innermost loop λ for individual I refers to the current loop $L_{\vec{x}_r}$ (lines 4, 10 and 17). Since λ denotes the position where a splitting-if would be inserted according to I, these lines of the fitness function determine whether loop nest splitting takes place in loop $L_{\vec{x}_r}$ or not. If this is not the case (lines 8, 14 and 21), the fitness function proceeds to the execution of the next nested loop $L_{\vec{x}_{r+1}}$.

In the other case, the fitness function exactly models the behavior of loop nest splitting in loop $L_{\vec{x}_r}$. For this purpose, an *if*-statement representing the corresponding splitting-if is introduced in the fitness function (lines 6, 12 and 19). Since the execution of *if*-statements has to be counted by the fitness function, the variable if_count is incremented by one beforehand (lines 5, 11 and 18).

This representative of a splitting-if simply has to determine if all the constraints contained only in the selected search space G_I are satisfied for the current values of the index variables $iv_1, \ldots, iv_r$. For this purpose, all constraints of a polytope R of G are connected with gene I_R using the logical *AND* operator. This is done for all polytopes R of G, and all these conditions for the polytopes R are connected with the logical *OR* (see the following example 5.13 for an illustration).

In the case that the splitting-if is not satisfied, the remaining loop nest $L_{\vec{\chi}_{r+1}}, \ldots, L_{\vec{\chi}_M}$ containing all original *if*-statements (compare figure 5.3) is executed. This behavior is modeled in the fitness function by the fact that the *if*-statements in lines 6, 12 and 19 check if G_I is not satisfied and then increment the counter if_count by $IF_{\lambda+1}$. Since a satisfied splitting-if implies that no *if*-statements are executed in the *then*-part of the splitting-if, the fitness function contains no code for the case that G_I is satisfied for $iv_1, \ldots, iv_r$.

The quotient $IF_1 /$ if_count is returned as final fitness value for an individual to the genetic algorithm so that the selection of the fittest individuals leads to the minimization of *if*-statement executions.

Let $iter_{max}$ denote the maximum number of iterations of a loop in a loop nest of depth N: $iter_{max} := \max\{ \lfloor (ub_i - lb_i) / \text{abs}(s_i) \rfloor + 1 \mid 1 \leq i \leq N \}$. Thus, the number of iterations executed by every loop of the fitness function is dominated by $iter_{max}$, and the number of loops contained in the fitness function is N in the worst case. For these reasons, it is simple to observe that the worst case runtime of this fitness function is dominated by the term $iter_{max}{}^N$ resulting in an exponential runtime complexity of the fitness function. The influence of this negative aspect is attenuated by the premise that this fitness function is used in order to optimize embedded data flow dominated source codes. In this situation, it can be assumed that the depth N of a loop nest typically is a small value. For example, the real-life applications studied in section 5.5 do not have a greater loop nest depth than seven. Additionally, the number of referenced loops L_{λ_r} which are modeled by the fitness function is typically much smaller than N as demonstrated by examples 5.11 and 5.13. Because of these reasons, it is not surprising that the actual runtimes of the genetic algorithm for global search space exploration are – with less than 0.4 CPU seconds – extremely low (see section 5.5).

EXAMPLE 5.13

For the global search space

$$G \;=\; \{x = 0 \land k = 0 \land y \geq 13\} \;\cup\; \{x \geq 10\} \;\cup$$
$$\{y = 0 \land 1 = 0 \land x \geq 9\} \;\cup\; \{y \geq 14\}$$

the following fitness function is used for global search space exploration:

```
double fitness(I₁, I₂, I₃, I₄) {
  int x, y, k, l, if_count = 0, λ = 0, tmp;

  for (tmp=1; tmp<=4; tmp++)
    if ((I_tmp == 1) && (λ_tmp > λ))
      λ = λ_tmp;

  for (x=0; x<=35; x++)
    if (λ == 2) {
      if_count++;
      if (!(I₂ && (x >= 10)))
        if_count += 127008;
    } else
    for (y=0; y<=48; y++)
      if (λ == 3) {
        if_count++;
        if (!((I₂ && (x >= 10)) ||
              (I₄ && (y >= 14))))
          if_count += 2592;
      } else
      for (k=0; k<=8; k++)
        if (λ == 4) {
          if_count++;
          if (!((I₁ && (x == 0) && (k == 0) && (y >= 13)) ||
                (I₂ && (x >= 10)) ||
                (I₄ && (y >= 14))))
            if_count += 288;
        } else
        for (l=0; l<=8; l++)
          if (λ == 5) {
            if_count++;
            if (!((I₁ && (x == 0) && (k == 0) && (y >= 13)) ||
                  (I₂ && (x >= 10)) ||
                  (I₃ && (y == 0) && (l == 0) && (x >= 9)) ||
                  (I₄ && (y >= 14))))
              if_count += 32;
          } else
          if_count += 32;
  return ((double) 1 / if_count);
}
```

In the beginning, λ is determined by checking all genes of I equal to 1 and by assigning the maximum value λ_r to λ. Under the assumption that the innermost loop λ for an individual I refers to the k-loop, the code within the then-part of if (λ == 4) is executed. Since λ equal to 4 implies that loop nest splitting is performed in the k-loop, the counter if_count is incremented by one in order to count the execution of the splitting-if.

The modeling of the splitting-if in the fitness function of this genetic algorithm is done by combining the genes of I with all constraints of the relevant regions R_r of G. Since λ refers to the k-loop, only the genes I_1, I_2 and I_4 can be equal to 1. Gene I_3 must be equal to 0, since otherwise λ would point to the l-loop. For this reason, I_3 is not checked by the if-statement

---

within the k-loop. *By combining the constraints of all regions* R_r *for which* I_r *can possibly be equal to 1 as illustrated, this if-statement models the selected search space* G_I.

If the conditions of this if-statement are not met for an individual I, the splitting-if would not be satisfied so that the remaining loop nest consisting of the l-, i- *and the* j-loop *together with the two if-statements within the* j-loop *would be executed. For this reason, the counter* if_count *is incremented by 288, since* $9 * 4 * 4 * 2$ *if-statements would be executed.*

After the termination of the genetic algorithm for global search space exploration, the selected search space $G_I = P_1 \cup \ldots \cup P_J$ is generated using the individual I having the highest fitness which has been created by the genetic algorithm. This finite union of polytopes G_I is the final solution of the entire optimization process.

The constraints of every polytope P_j of G_I define ranges of values for the index variables of Λ where all loop-variant *if*-statements are satisfied. Furthermore, G_I has been generated in such a way that the number of *if*-statement executions is always minimized. Due to these reasons, the finite union of polytopes $G_I = P_1 \cup \ldots \cup P_J$ exactly represents the set of alternatives $\alpha = \{A_1, \ldots, A_J\}$ which fulfills all optimization criteria for loop nest splitting as specified in definition 5.2.

5.3.6   Source Code Transformation

In a final step, the finite union of polytopes $G_I = P_1, \ldots, P_J$ created during global search space exploration is used to transform the code of the loop nest $\Lambda = \{L_1, \ldots, L_N\}$. The transformation of Λ is done in two steps:

1 The conditions of the splitting-if have to be generated according to the constraints contained in G_I.

2 Parts of the loop nest Λ have to be duplicated and modified in order to form the *then-* and *else*-parts of the splitting-if.

5.3.6.1   Generation of the splitting If-Statement

In order to explain the way how the conditions of the splitting *if*-statement are created out of a finite union of polytopes G_I, the structure of the constraints contained in G_I needs to be clarified.

THEOREM 5.3 (STRUCTURE OF CONSTRAINTS IN G_I)
Let $\Lambda = \{L_1, \ldots, L_N\}$ *be a loop nest and* $G_I = P_1 \cup \ldots \cup P_J$ *be the result of global search space exploration where every polytope of* G_I *is defined by a set of equations and inequations:* $P_j = \{\, i \in \mathbb{Z}^N \mid Ai = a, \ Bi \geq b \,\}$.
Every constraint in P_j, $1 \leq j \leq J$, *has the structure*

$$i_l = c_1 \ \text{ or } \ i_l \geq c_2$$

for an index variable $i_l \in \mathbb{Z}$ *and constant values* $c_1, c_2 \in \mathbb{Z}$.

Proof

The initial point leading to the structure of G_I are the polytopes P_C generated during the step of condition optimization. According to definition 5.9, P_C only contains constraints like $lb_l \leq i_l$, $i_l \leq ub_l$, $i_l \geq v'_{C,l}$ or $i_l \leq v'_{C,l}$, where lb_l, ub_l and $v'_{C,l}$ are constants. Since $i_l \leq c \Leftrightarrow -i_l \geq -c$, the constraints of P_C are compliant with theorem 5.3.

The polytopes P_C are combined with each other using the union and intersection operators. By themselves, these operators do not modify the structure of the constraints. If a union of polytopes is performed, the two polytopes are combined to a new finite union of polytopes without touching the constraints. In the case of an intersection, all constraints of both polytopes are merged into a single resulting polytope.

But as mentioned in section 3.1, the extended Motzkin algorithm [Wilde, 1993] is applied implicitly to each resulting polytope in order to minimize the set of constraints. In a first step, a Gaussian elimination of variables is performed. Since by definition all constraints of P_C depend on exactly one variable, this elimination leads to the entire removal of constraints, whereas the remaining constraints are not modified. In a second step, redundancies between constraints are detected. If one constraint defines a superset of another one, the former constraint can be removed completely. If two constraints $a \geq c$ and $a \leq c$ exist, they are coalesced to $a = c$. In all cases, all constraints still depend on exactly one variable and a constant value so that it can be concluded that the global search space G as defined in section 5.3.4 meets the properties of the above theorem.

Since the finite union of polytopes G_I using the fittest individual I after search space exploration is a subset of the global search space G, theorem 5.3 holds for G_I.

$\square$

Theorem 5.3 is an aid for distinguishing between constraints in G_I containing information that was obtained by the genetic algorithms on the one hand and constraints modeling only the loop bounds on the other hand. Since the adherence of lower and upper loop bounds is explicitly guaranteed by the loop nest Λ itself, the splitting-if does not need to contain corresponding conditions. Rather, only the important information gathered during condition optimization and global search space exploration needs to be expressed by the splitting-if. For this purpose, the notion of *relevant constraints* is introduced as follows:

DEFINITION 5.16 (RELEVANT CONSTRAINTS)
Let $\Lambda = \{L_1, \ldots, L_N\}$ be a loop nest and the finite union of polytopes G_I be the result of global search space exploration having the structure presented in theorem 5.3.

*a A constraint $i_l \geq c_1$ or $-i_l \geq -c_2$ of G_I is **irrelevant** if c_1 is equal to lb_l or c_2 is equal to ub_l respectively.*

*b All constraints of G_I which are not irrelevant are said to be **relevant**.*

Using the relevant constraints of G_I, the conditions of the splitting *if*-statement are constructed in the following way:

DEFINITION 5.17 (SPLITTING IF-STATEMENT)
Let $G_I = P_1 \cup \ldots \cup P_J$ be the result of global search space exploration and let $C_{j,1}, C_{j,2}, \ldots$ denote the relevant constraints of polytope P_j of G_I.

*The **splitting-if** is composed of these relevant constraints by translating them into ANSI-C syntax in a straightforward way and by connecting them as follows:*

```
if ((C_{1,1} && C_{1,2} && ...) || (C_{2,1} && C_{2,2} && ...) || ... ||
    (C_{J,1} && C_{J,2} && ...))
```

As can be seen, the splitting *if*-statement defined above directly models the structure of G_I which has been shown to fulfill the optimization criteria for loop nest splitting presented in section 5.2. For this reason, loop nest splitting using a splitting-if according to definition 5.17 guarantees that all *if*-statements in Λ are satisfied in the case that the splitting-if itself is satisfied. Additionally, this structure of the splitting-if is a very compact and efficient way to specify the iteration ranges where all *if*-statements are satisfied under simultaneous minimization of *if*-statement executions. This property is due to the fact that only relevant constraints which do not contain any redundancies because of the Motzkin algorithm are considered for the formulation of the splitting-if.

5.3.6.2   Loop Nest Duplication

The fundamental property of loop nest splitting that the satisfaction of the splitting-if implies that all loop-depending *if*-statements in loop nest $\Lambda = \{L_1, \ldots, L_N\}$ are also satisfied has to be reflected by the *then*- and *else*-parts of the splitting-if. In order to generate code for these *then*- and *else*-parts, parts of the original loop nest Λ have to be duplicated.

For a finite union of polytopes G_I being the result of the entire optimization process, λ denotes the index of the innermost loop where the loop nest has to be split (see definition 5.14d). As a consequence, the splitting-if has to be placed in loop L_λ.

The *else*-part of the splitting-if denoted as EP_λ can be created easily, because it mainly consists of a direct copy of the body of loop L_λ. Since the *else*-part is executed only when the splitting-if is not satisfied, the original *if*-statements located in EP_λ may be satisfied or not. In this situation, nothing can be said about the satisfaction of these original *if*-statements so that they have to be kept in the *else*-part of the splitting-if in order to guarantee the correctness of the transformation.

The only subject to optimization in EP_λ are those conditions of original *if*-statements which have been proven to be satisfied or unsatisfied for all iterations of the loop nest. This information is collected during the step of condition satisfiability (see section 5.3.2). Each condition which is constantly satisfied for all iterations of Λ is removed from its *if*-statement in the *else*-part of the splitting-if and is replaced by the constant "1" (unsatisfiable conditions analogously).

The *then*-part TP_λ of the splitting-if is also based on a copy of the body of loop L_λ. Since TP_λ is executed only when the splitting-if is satisfied, TP_λ does not need to contain any original *if*-statement since it is known that they are satisfied also. For this reason, TP_λ is traversed recursively. For every occurrence of a loop-variant *if*-statement in TP_λ, only its *then*-part is kept in TP_λ. The loop-variant *if*-statement itself including all its conditions and its possibly existing *else*-part are removed completely from TP_λ.

In the case that the conditions of the splitting-if define a single unique range $[lo_\lambda, hi_\lambda]$ of iterations for the innermost loop L_λ where all *if*-statements are satisfied, TP_λ can be surrounded by a new *for*-loop as shown in the example given in figure 5.3 (see page 59). This surrounding *for*-loop modifies the index variable of the innermost loop i_λ in such a way that it steps from lo_λ to hi_λ using the given loop stride s_λ. This surrounding *for*-loop has the advantage that the range of iterations lo_λ, hi_λ for the innermost loop L_λ is traversed without executing the splitting-if at all, leading to an additional reduction of *if*-statement executions.

EXAMPLE 5.14

For the loop nest depicted in figure 5.1, the analysis techniques presented in this chapter lead to the result that all if-statements are satisfied and that the if-statement executions are minimized if the following splitting-if is used:

$$\text{if (x >= 10 || y >= 14)}$$

For the innermost y-loop, the conditions of the splitting-if and the loop bounds imply that the splitting-if is always satisfied for the range $14 \leq y \leq 48$. As a consequence, the then-*part of the splitting-if contains the surrounding loop* for (; y<49; y++) *(see figure 5.3). Since this surrounding loop is executed only if the splitting-if is satisfied, it is ensured that y is already greater than or equal to 14. For this reason, the explicit formulation of a lower bound for the surrounding loop can be omitted leading to a compact source code after the transformation.*

Assuming that the analysis for loop nest splitting leads to the following splitting-if

$$\text{if ((x >= 10 \&\& x <= 20 \&\& y >= 14 \&\& y <= 42) ||}$$
$$\text{(x >= 25 \&\& x <= 35 \&\& y >= 17))}$$

a surrounding y-loop is not generated in the then-*part of this splitting-if. This is because of the fact that the conditions on the left side of the || operator specify the iteration range $14 \leq y \leq 42$, whereas the conditions on the right hand side imply $17 \leq y \leq 48$. As a consequence, the surrounding y-loop could only be allowed to count to the upper bound of 42. In contrast, the remaining iterations $43 \leq y \leq 48$ would have to be treated by particular* *code leading to the insertion of additional if-statements in order to detect this particular range of iterations. This situation is highly undesirable due to the additional code size increases and the counterproductive modification of the control flow. For these reasons, the generation of the surrounding y-loop is suppressed in the absence of a common upper bound hi_λ.*

After the complete generation of the splitting *if*-statement consisting of its *then*-part TP_λ – possibly surrounded by a new *for*-loop – and its *else*-part EP_λ, the body of loop L_λ is removed completely from the loop nest Λ. Instead of the removed code, the splitting-if is inserted. This replacement in loop nest Λ leads to the transition from the code depicted in figure 5.1 to the one illustrated in figure 5.3 after loop nest splitting.

5.4   Extensions for Loops with non-constant Bounds

As shown by the experimental results given in section 5.5, the approach for loop nest splitting presented in the previous section 5.3 is already highly effective. Nevertheless, it is still restricted to very particular classes of loop nests due to the precondition (see section 5.3.1) that all lower and upper bounds lb_l and ub_l of a loop L_l be constant values. Because of this restriction, the previously presented loop nest splitting could not be applied to the code fragment depicted in figure 5.9, since the upper bound of the innermost j-loop depends on the outer index variable i. This section presents extensions to the previously explained analysis techniques which eliminate this restriction and allow loop nest splitting to be applied to more general classes of embedded software including frequently occurring sorting algorithms and DSP filter routines.

The basic structure of the analysis algorithms for loop nest splitting consisting of the verification of the condition satisfiability, condition optimization, global search space construction and global search space exploration remains unchanged even after the extensions proposed in the following. The main advantage of constant loop bounds can be seen in the simplicity of the genetic algorithm for condition optimization. For a set of values $v'_{C,l}$ generated by this genetic algorithm, the fitness function only needs to evaluate some expressions mainly consisting of sum-of-products of the constant loop bounds and the $v'_{C,l}$ values. Using these formulas, the exact number of *if*-statement executions can be computed and minimized.

As already mentioned in the previous section, polyhedral models are an integral part of the analysis for loop nest splitting. Since these models base on linear inequations and equations by definition, loop bounds are allowed to be affine expressions in the following:

DEFINITION 5.18 (AFFINE LOOP BOUNDS)
Let $\Lambda = \{L_1, \ldots, L_N\}$ be a loop nest of depth N.

```
for (i=0; i<50; i++)
    for (j=0; j<i; j++) {
        if (i+j<70)
            then_block;
        else
            else_block;
            ⋮
    }
```

Figure 5.9. A typical Loop Nest with variable Loop Bounds

a *For the outermost loop L_1, the lower and upper bounds lb_1 and ub_1 are constant values.*

b *For any other loop $L_l \in \{L_2, \ldots, L_N\}$, the lower and upper **bounds** can be **affine** expressions of the surrounding index variables $i_1, \ldots, i_{l-1}$. As a consequence, the index variable i_l iterates between*

$$lb_l = \sum_{j=1}^{l-1} (c'_j * i_j) + c' \le i_l \le \sum_{j=1}^{l-1} (c''_j * i_j) + c'' = ub_l$$

for constants $c'_j, c', c''_j, c'' \in \mathbb{Z}$.

The fact that the outermost loop is required to have constant loop bounds ensures that a loop nest is still fully analyzable at compile time due to the absence of external data dependencies[4]. On the other hand, the variability of the inner loop bounds has the effect that the number of *if*-statement executions minimized during condition optimization not only depends on the values $v'_{C,l}$ generated by the genetic algorithm. In addition, implicit dependencies on the value of an index variable i_j exist, since i_j possibly influences the bounds of an inner loop L_l. Due to these complications, no way has been found to allow the fitness function of the genetic algorithm to count *if*-statement executions using simple numerical formulas.

One of the main ideas of the genetic algorithm for condition optimization presented in section 5.3.3 is the combination of the monotony of conditions (see theorem 5.1) with constant loop bounds. This property allows the use of a very efficient chromosomal encoding, because the determination of a pair of values $(lb'_{C,l}, ub'_{C,l})$ for every loop L_l can be achieved using one gene representing $v'_{C,l}$, and because invalid individuals are easy to detect. Assuming that affine loop bounds as defined above are allowed in the following, a similar genetic

---

[4]The integration of data flow analysis techniques into the loop nest splitting framework is possible, but is beyond the scope of this book.

encoding consisting only of a value $v'_{C,l}$ for every loop L_l can not be used. This is because it can not be concluded that a condition C is satisfied for the entire interval $[v'_{C,l}, ub_l]$ (or $[lb_l, v'_{C,l}]$) since the truth value of C may now depend on outer loops. Additionally, the designation of the innermost loop λ can not be done using only the intervals $[lb'_{C,l}, ub'_{C,l}]$, because these intervals have to be true sub-intervals of the loop bounds (see definition 5.7c) which are potentially variable.

Due to these reasons, the following chromosomal encoding is used in order to optimize conditions under consideration of affine loop bounds:

DEFINITION 5.19 (CHROMOSOMAL REPRESENTATION)
Let $\Lambda = \{L_1, \ldots, L_N\}$ *be a loop nest of depth* N *with affine lower and upper loop bounds,* $C = \sum_{l=1}^{N}(c_l * i_l) \geq c$ *a condition.*

a *A **chromosome** is an array of integer values of length* $2 * N + 1$.

b *For* $l \in \{1, \ldots, N\}$ ***gene** number* $2 * l - 1$ *denotes the value* $lb'_{C,l}$, *whereas gene* $2 * l$ *represents* $ub'_{C,l}$.

  Using these genes, the genetic algorithm for condition optimization specifies the intervals $[lb'_{C,l}, ub'_{C,l}]$ *for every loop* L_l *so that condition* C *is assumed to be satisfied for all values of the index variable* i_l *within this interval.*

c *The last gene* $2 * N + 1$ *stores the index of the **innermost loop*** λ *for loop nest splitting.*

  The interpretation of this gene is that loop nest Λ *has to be split at loop* L_λ, *i. e. the splitting-if will be placed in loop* L_λ.

The fitness function supporting affine loop bounds which is presented in the following is structured according to that used for global search space exploration (see section 5.3.5). It models an entire loop nest after splitting using the information stored in a chromosome as specified by definition 5.19. This new structure of the fitness function for the optimization of a condition C is depicted as pseudo-code in figure 5.10.

The main goal of this function is to update two counters in the correct way. The first one (`if_count`) stores the number of *if*-statement executions for a given individual I. The variable `penalty` counts how many times condition C is not satisfied when it is supposed by I that it should be satisfied. This latter counter serves for the detection of individuals generated by the genetic algorithms representing illegal solutions.

Principally, this fitness function contains the entire loop nest Λ as can be seen from lines 3 and 19 of figure 5.10 (the dots in line 18 denote the omitted loops $L_2, \ldots, L_{N-1}$). For every loop, some code is required for modeling the

```
1   double fitness(individual I) {
2       int if_count = 0, penalty = 0;

3       for (i_1=lb_1; i_1<=ub_1; i_1+=s_1)
4           if (I_{2*N+1} == 1) {
5               if_count++;
6               if ((i_1 >= I_1) && (i_1 <= I_2)) {
7                   for (; i_1<=I_2; i_1+=s_1)
8                       for(i_2=lb_2; i_2<=ub_2; i_2+=s_2)
9                           ...
10                          for (i_N=lb_N; i_N<=ub_N; i_N+=s_N)
11                              if (!satisfied(C)) penalty++;
12              } else
13                  for (i_2=lb_2; i_2<=ub_2; i_2+=s_2)
14                      ...
15                      for (i_N=lb_N; i_N<=ub_N; i_N+=s_N)
16                          if_count++;
17          } else

18              ⋮

19          for (i_N=lb_N; i_N<=ub_N; i_N+=s_N)
20              if (I_{2*N+1} == N) {
21                  if_count++;
22                  if ((i_1 >= I_1) && (i_1 <= I_2) && ... &&
                        (i_N >= I_{2*N-1}) && (i_N <= I_{2*N})) {
23                      for (; i_N<=I_{2*N}; i_N+=s_N)
24                          if (!satisfied(C)) penalty++;
25                  } else
26                      if_count++;
27              }

28      if (penalty == 0) return ((double) (IS_Λ / if_count));
29      else return ((double) (IS_Λ / (penalty + ERR))); }
```

Figure 5.10. Structure of the Fitness Function for Condition Optimization

loop nest after a potential splitting (lines $4-17$ resp. $20-27$). When entering a loop L_l, it is first checked whether this loop contains the splitting-if. For this purpose, the gene of I storing the index of the innermost loop λ is referenced (lines 4 and 20). If this is not the case, the algorithm proceeds to loop L_{l+1} (line 17).

Otherwise, loop L_l contains the splitting-if (lines 6 and 22) whose execution requires an increment of if_count (lines 5 and 21). The goal of the splitting-if for loop L_l is to check the genes $I_1, \ldots, I_{2*l}$ and to verify that the index variables $i_1, \ldots, i_l$ actually are within the intervals specified by these genes. If the splitting-if is true, the duplicated loop L_l counting to the upper bound I_{2*l} of the range is executed next (lines 7 and 23). Within this loop, the remaining loop nest $L_{l+1}, \ldots, L_N$ can be found (lines 8 – 10). Since the splitting-if is true when executing this code, it is assumed that condition C also is true, so that the counter if_count remains unchanged in lines 7 – 11 resp. 23 – 24. But since the genetic algorithm can generate invalid individuals where the satisfaction of the splitting-if does not imply that condition C is true, care has to be taken in order to detect these situations. Lines 11 and 24 check whether the condition C is true or not. If it is false, an invalid individual I is detected and the counter penalty is incremented.

Finally, some code is required for the case that the splitting-if in a loop L_l is not true. In analogy to figure 5.3, the remaining loop nest $L_{l+1}, \ldots, L_N$ is contained in the *else*-part of the splitting-if (see lines 13 – 15). In loop L_N, the counter if_count is incremented (lines 16 and 26) since the *else*-part of the splitting-if necessarily contains an *if*-statement for checking condition C.

After the execution of the entire loop nest depicted in lines 3 – 27, the fitness of an individual I is calculated based on the values of if_count and penalty. If an individual represents a valid solution, counter penalty is equal to zero (line 28). In this case, the quotient of IS_Λ and if_count is returned as fitness value ensuring again that individuals leading to fewer *if*-statement executions have a higher fitness. In the case of an invalid individual (line 29), a large constant *ERR* is added to penalty, and the quotient of this sum is returned to the genetic algorithm. By carefully selecting *ERR*, it can be guaranteed that illegal individuals can never have a better fitness than legal ones. This is done by setting *ERR* to the size of the total iteration space of Λ, since if_count can not be incremented more often than loop iterations are executed.

As can be seen from figure 5.10, the runtime of the fitness function depends on the depth N of a loop nest and on the actual affine loop bounds. For this reason, the same argumentation concerning the complexity of the fitness function holds as already discussed in the context of global search space exploration (see section 5.3.5).

After the termination of this new genetic algorithm for condition optimization, the best individual I_{Opt} having the maximum fitness is used in order to generate a polytope P_C for condition C. This step is very similar to the technique described in section 5.3.3.3, because here again, the intervals represented by I_{Opt} have to be modeled properly by P_C. For this purpose, P_C consists of four constraints for every loop L_l. Two constraints are used for the representation of the lower and upper loop bounds, and the other two constraints model the interval $[lb'_{C,l}, ub'_{C,l}]$ encoded in I_{Opt}. Formally, P_C is specified by the following definition:

DEFINITION 5.20 (OPTIMIZED POLYTOPE)
Let $\Lambda = \{L_1, \ldots, L_N\}$ be a loop nest with affine lower and upper bounds and C be a condition. $lb'_{C,l}$ resp. $ub'_{C,l}$ denote the values stored in the genes of the individual I_{Opt} with maximum fitness for a loop $L_l \in \Lambda$.

*The **polytope P_C** modeling the optimized iteration ranges specified by I_{Opt} is defined as follows:*

$$P_C = \{ (x_1, \ldots, x_N) \in \mathbb{Z}^N \mid \forall L_l \in \Lambda : lb_l \leq x_l \leq ub_l \wedge$$
$$x_l \geq lb'_{C,l} \wedge x_l \leq ub'_{C,l} \}$$

The optimization step performed immediately after condition optimization deals with the construction of the global search space (compare section 5.3.4). Since in this phase, only the polytopes P_C generated during condition optimization are connected using the union and intersection operators for finite unions of polytopes, no modifications have to be applied to the techniques used during search space construction in order to model affine loop bounds.

For the final step of global search space exploration, only slight changes have to be applied to the techniques presented in section 5.3.5. As stated in the description of the genetic algorithm for search space exploration, the fitness function only consists of the relevant loops, i. e. those loops L_l which are explicit candidates for loop nest splitting when considering a global search space G. In a situation where affine loop bounds are allowed, the fitness function can not only consist of those relevant loops, because the bounds of a relevant loop L_l can depend on an outer loop which will never contain the splitting-if. For this reason, all those non-relevant loops that the relevant loops depend on have to be added to the fitness function. Using this extended fitness function, the genetic algorithm for global search space exploration leads to a resulting selected search space G_I as described previously.

In order to transform a loop nest and to generate the conditions of the splitting *if*-statement after the entire optimization process, a theorem describing the structure of constraints contained in the final finite union of polytopes G_I was used (see theorem 5.3 in section 5.3.6). This theorem states that every constraint of G_I has the shape $i_l = c_1$ or $i_l \geq c_2$ for constant values c_1 or c_2 respectively. In the case of affine loop bounds, it is obvious that this theorem is invalid, because constraints of G_I modeling the loop bounds can be affine expressions. As a consequence, the term of relevant constraints introduced in definition 5.16 specifying those constraints that have to be modeled by explicit conditions in the splitting-if has to be re-formulated:

DEFINITION 5.21 (RELEVANT CONSTRAINTS)
Let $\Lambda = \{L_1, \ldots, L_N\}$ be a loop nest and G_I be the result of global search space exploration.

a *A constraint $i_l \geq lb_l$ (or $-i_l \geq -ub_l$) of G_I is **irrelevant** if loop L_l has a constant lower loop bound of lb_l (or ub_l resp.).*

b *A constraint $\sum_{j=1}^{N}(c_j * i_j) \geq c$ with constants c_j and c is **irrelevant** if a loop L_l*

   with affine bounds exists where the constants c'_j and c' (or c''_j and c'' resp.) introduced in definition 5.18 are pairwise equal to c_j and c.

c *All constraints of G_I which are not irrelevant are said to be **relevant**.*

The irrelevant constraints specified in definition 5.21 consist of all those constraints modeling nothing but loop bounds. Item 5.21a obviously defines constant loop bounds, whereas item 5.21b is a reformulation of affine loop bounds. All other constraints not being a loop bound contain relevant information. Using this notion of relevant constraints, the splitting-if can be generated in accordance to definition 5.17, so that the process of source code transformation can be performed as described in section 5.3.6 without any further changes.

The application of the techniques presented in this section to the code depicted previously in figure 5.9 leads to the loop nest shown in figure 5.11 and described in the following example.

EXAMPLE 5.15

The analysis for loop nest splitting observes that the outer i-loop iterates from 0 to 49, while the inner j-loop steps from 0 to the actual value of i. Considering the condition i+j<70, the techniques presented in this section recognize that this condition must necessarily be true for all values of i smaller than or equal to 35. Using this information, the loop nest is split as illustrated in figure 5.11 leading to a reduction of if-statement executions from 1,125 down to 610.

5.5   Experimental Results

All algorithms presented in sections 5.3 and 5.4 are fully implemented and integrated into a single tool for source code transformation. As mentioned in chapters 4 and 3, this loop nest splitting tool is based on the SUIF intermediate representation [Wilson et al., 1995] and the Polylib [Loechner, 1999] and PGAPack [Levine, 1996] libraries.

In this section, the impacts of this optimization on pipelines and caches, runtimes, code sizes and energy dissipation are presented in detail. The development of loop nest splitting was initially motivated by the control flow overhead implicitly contained in data dominated software on the one hand,

```
for (i=0; i<50; i++)
  if (i <= 35)
    for (; i<=35; i++)
      for (j=0; j<i; j++) {
        then_part;
          ⋮
      }
  else
    for (j=0; j<i; j++) {
      if (i+j<70)
        then_part;
      else
        else_part;
          ⋮
    }
```

Figure 5.11. A Loop Nest with variable Loop Bounds after Splitting

and by data layout optimizations explicitly inserting control flow overhead on the other hand (cf. sections 5.1.1 and 5.1.2). According to these two aspects of motivation, the presentation of the results obtained during benchmarking is structured. In section 5.5.1, the efficacy of stand-alone loop nest splitting is founded by applying the optimization to a set of three benchmarks and by demonstrating to what extent control flow bottlenecks are eliminated from these benchmarks. Section 5.5.2 presents detailed experimental results for a framework where loop nest splitting is applied in combination with a data partitioning optimization.

5.5.1 Stand-alone Loop Nest Splitting

The tool for loop nest splitting was applied to the source codes of three multimedia programs from the image processing domain. The first one is a medical image processing application (*CAVITY* [Bister et al., 1989]) which extracts contours from computer tomography images to help physicians detect brain tumors. Since this application served as test driver for the so called DTSE transformations presented in section 2.3, it is used here as benchmark in order to show that loop nest splitting is able to remove the overhead introduced by DTSE. The cavity detector consists of two nested loops scanning an image in its x and y dimensions. The index variables iterate from 0 to 1,002 (y) and 0 to 1,282 (x) leading to a total amount of 1,286,849 loop iterations. During each iteration, 13 *if*-statements suitable for loop nest splitting are executed containing 40 index variable accesses and 71 arithmetical and / or logical ANSI-C operations.

	CAVITY	ME	QSDPCM
Nest Depth	2	7	7
# Iterations	1,286,849	45,722,880	12,830,400
# *If*-Statements	13	2	1
# Variable Accesses	40	8	12
# Operations	71	14	19
Optimization Runtime	0.37 s	0.16 s	0.08 s

Table 5.1.   Characteristic Properties of Benchmarks

The second benchmark is the MPEG 4 full search motion estimation kernel (*ME* [Gupta et al., 2000]) which already served as example throughout this chapter. From figure 5.1 it can be seen that the sevenfold loop nest defines an iteration space of size 45,722,880. The innermost loop of the ME benchmark contains two *if*-statements with a total of eight index variable accesses and 14 ANSI-C operations.

Finally, the QSDPCM algorithm [Strobach, 1988] for scene adaptive coding serves as third test driver. Like the previous benchmark, its loop nest has a depth of seven; a total of 12,830,400 iterations of the innermost loop are executed. In the hot-spot of this application, one *if*-statement containing 12 index variable references and 19 arithmetical / logical operations can be found.

The main characteristics of the benchmarks are summarized in table 5.1. Since all lower and upper loop bounds of these benchmarks are constant, loop nest splitting solely consists of the optimization algorithms presented in section 5.3. The runtimes of the loop nest splitting tool are very low. For the optimization of CAVITY, including the polyhedral and genetic algorithms as well as the transformation of the code, on an Intel Pentium IV (2.6 GHz), 0.37 CPU seconds are required (ME: 0.16 s, QSDPCM: 0.08 s). In order to generate all results presented in this section, the techniques described in section 3.3 were applied. Detailed lists of the data gathered during benchmarking are contained in appendix B.

5.5.1.1   Pipeline and Cache Performance

In this section, the effects of loop nest splitting on several cache and pipeline parameters of three different processors are demonstrated. Because of their easy to access on-chip debugging and profiling facilities, the results reported in the following were measured with an Intel Pentium III processor, a Sun UltraSPARC III and a MIPS R10000. Short descriptions of the architectures of

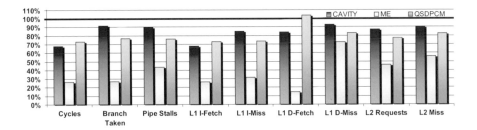

Figure 5.12. Relative Pipeline and Cache Behavior for Intel Pentium III

these CPUs and of the tools used for performance monitoring can be found in appendix B.1.

Intel Pentium III

Figure 5.12 shows how the instruction pipeline, the caches and the transfers of data and instructions on the system buses are affected by loop nest splitting. The results for the optimized benchmarks after loop nest splitting are given as a percentage of the unoptimized versions which are denoted as 100% in the diagram.

The most important observation illustrated in figure 5.12 is that loop nest splitting leads to significant speed-ups of the benchmarks using the Intel Pentium III processor (see column Cycles). After the application of the source code optimization presented in this chapter, runtimes decrease between 26.7% in the case of QSDPCM and up to 73.5% for ME, thus clearly demonstrating the high effectiveness of this transformation. In the following, the individual factors contributing to these speed-ups are analyzed and presented in detail.

As can be seen from the columns Branch Taken and Pipe Stalls, loop nest splitting is able to generate a more regular control flow for all three benchmarks. The number of taken branch instructions is reduced between 8.1% (CAVITY) and 72.9% (ME) consequently leading to similar reductions of pipeline stalls (10.4% – 56.5%). These numbers underline that even in the domain of data dominated multimedia software, control flow issues should not be neglected since they can cause significant penalties to an instruction pipeline.

The fact that loop nest splitting removes control flow overhead and thereby significantly improves the runtime behavior of instruction caches is reflected by the columns L1 I-Fetch and L1 I-Miss. As can be seen, the total number of instruction fetches is reduced by 26.7% (QSDPCM) resp. 73.6% (ME), and the total amount of instruction cache misses is improved between 14.8% (CAVITY) and 68.5% (ME).

Due to the removal of conditions and the involved reduction of index variable accesses, the L1 data cache also benefits from the optimization. Data fetches from this cache (column L1 D-Fetch) are reduced by 16% (CAVITY) resp. 85.4% (ME). Only in the case of the QSDPCM benchmark, the amount of data fetches increases by 3.9% due to the insertion of spill code by the compiler. Nevertheless, this slight increase is negligible since all other performance related parameters for this benchmark show large improvements. In addition, the absolute amount of data cache misses depicted in column L1 D-Miss of figure 5.12 drops by 7.2% (CAVITY) up to 27.2% (ME).

Transfers of instructions and data between L1 caches, the unified L2 cache and main memory are also optimized by loop nest splitting. As can be seen from column L2 Requests, the L2 cache is referenced 13.1% (CAVITY) up to 53.8% (ME) less than before the optimization. When focusing on the system bus connecting the Pentium with its main memory, between 9.9% (CAVITY) and 43.8% (ME) less load can be observed (column L2 Miss).

Sun UltraSPARC III

Figure 5.13 shows the effects of loop nest splitting on the instruction pipeline and the caches of the UltraSPARC III processor. Compared to the Intel Pentium CPU, even higher gains with respect to the runtimes of the benchmarks were achieved. For this environment, speed-ups ranging from 33.3% (CAVITY) up to 75.8% (ME) were measured.

In the case of the Sun UltraSPARC processor, these speed-ups are mainly due to the improvement of the instruction pipeline behavior after loop nest splitting as can be seen from figure 5.13. This architecture has a very long and complex instruction pipeline consisting of fourteen stages making it very sensitive to stalls. For this reason, loop nest splitting leads to very high gains for the SPARC processor (see columns Branch Taken and Pipe Stalls). The total number of pipeline stalls is reduced by 29.1% in the case of the CAVITY benchmark. For the QSDPCM application, improvements of 55.5% were observed, and the optimization of the ME benchmark leads to 73.1% less pipeline stalls.

When considering the number of taken branches as an indicator of the linearity of the control flow, similar improvements can be observed. After loop nest splitting, the execution of the QSDPCM benchmark leads to 54% less branches, in the case of the MPEG-4 motion estimation, a reduction by 88.3% can be reported. The fact that the branching behavior of the CAVITY detector becomes worse by 5.3% after the optimization is remarkable. Further experiments and measurements have shown that this effect is caused by the current version (6.2) of the Sun Workshop compiler used for benchmarking, and that this is not a misconduct of loop nest splitting itself:

In the code of the CAVITY benchmark, a sequence of eight *if*-statements checking data-dependent conditions can be found. These *if*-statements are not

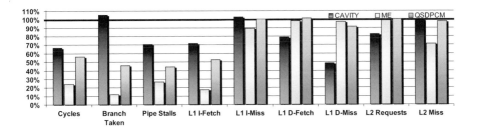

Figure 5.13. Relative Pipeline and Cache Behavior for Sun UltraSPARC III

suitable for loop nest splitting. Therefore, they are duplicated by the source code transformation – one time in the *then*-part of the splitting-if, and one time in the *else*-part. It has turned out that these *if*-statements are the reason for the increase of taken branches when using the Sun compiler. When commenting out these data-dependent *if*-statements, the CAVITY detector only contains *if*-statements that can be processed by loop nest splitting. Under these circumstances, the branching behavior of the benchmark is improved as expected by 28.4%. This number shows that the gains achieved by the optimization of loop-variant *if*-statements are compensated by some particular transformations of the used version of the Sun compiler applied to the data-dependent *if*-statements. The assumption that this current compiler version is the reason for the observed behavior can be validated by using the previous version 6.1 instead. In this case, the profiling of CAVITY including the data-dependent *if*-statements leads to a reduction of 33.7% of taken branches.

As in the case of the Pentium processor, the number of accesses to the L1 instruction cache is reduced significantly. The improvements vary between 28.2% (CAVITY) and 82.7% (ME). In contrast, the amount of I-cache misses is only reduced for the ME benchmark (10.4% improvement). For the other benchmarks, the number of cache misses remains nearly unchanged (less than 0.1% of improvement for QSDPCM and 2.6% degradation for CAVITY).

When focusing on the level 1 data cache, it can be seen that the number of accesses to it is only reduced after the optimization of the CAVITY benchmark. In this case, 21.2% less data transfers concerning this cache are performed. For the other benchmarks, the number of data fetches stays constant. The reason for this behavior is the large register file of the UltraSPARC processor containing 160 general purpose integer registers out of which 24 are accessible without limitations at any point of time. Since the ME and QSDPCM benchmarks only use very few local variables, the compiler is able to store them entirely

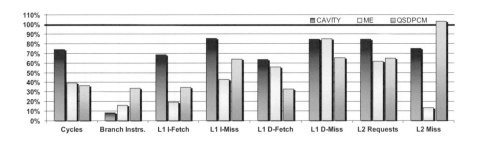

Figure 5.14. Relative Pipeline and Cache Behavior for MIPS R10000

in processor registers even before loop nest splitting. The number of L1 data cache misses is reduced between 2.9% (ME) and 51.4% (CAVITY).

As can be seen from the columns L2 Requests and L2 Miss, the level 2 unified cache also benefits from loop nest splitting. In the case of the CAVITY benchmark, the number of accesses to this cache is reduced by 17.2%, whereas no changes were observed for the other benchmarks. L2 cache misses are only reduced for the ME benchmark (improvement of 28.6%), the transformation of the other applications only leads to marginal improvements (1.1% for CAVITY, 2.3% for QSDPCM).

MIPS R10000

The behavior of the performance counters of the MIPS R10000 processor after loop nest splitting is depicted in figure 5.14. As illustrated in column Cycles, loop nest splitting also leads to accelerations of the benchmarks for this architecture. Actually, the runtimes are reduced between 26.1% (CAVITY) up to 63.4% (QSDPCM).

Unfortunately, the MIPS performance counters are not able to profile the behavior of the instruction pipeline so that no detailed data with respect to pipeline stalls and branching behavior can be given in this section. Only the total number of executed branch instructions can be measured. The column Branch Instrs of figure 5.14 denotes all branch instructions – taken branches as well as untaken jumps – evaluated during the execution of the benchmarks. This column shows that reductions of executed branch instructions between 66.3% (QSDPCM) and 91.8% (CAVITY) were achieved by loop nest splitting.

The MIPS caches benefit from the source code optimization to a large extent. As can be seen in figure 5.14, loop nest splitting is able to reduce the number of accesses to the level one caches significantly. Instruction fetches are decreased between 31.5% (CAVITY) and 81% (ME), whereas data cache accesses are alleviated between 36.4% (CAVITY) and 67.3% (QSDPCM). Due to these

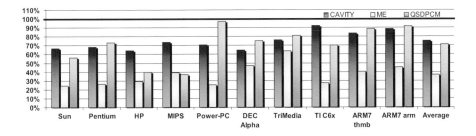

Figure 5.15. Relative Runtimes after Loop Nest Splitting

improvements of cache accesses, the number of cache misses is also reduced. The number of level 1 instruction cache misses is diminished between 14.7% (CAVITY) and 56.9% (ME), data cache misses occur between 15.2% (ME) and 34.8% (QSDPCM) less frequently after the optimization.

Improvements related to the level 2 unified cache are observed likewise. The number of read / write requests to this cache is reduced between 15.4% (CAVITY) and 38.3% (ME). Level 2 cache misses are slightly increased by 3.3% for the QSDPCM benchmark and are decreased by 25% (CAVITY) and 87% (ME).

5.5.1.2 Execution Times and Code Sizes

In this section, it is shown that the improvements reported previously in section 5.5.1.1 are not restricted to the platforms presented there. A large collection of different processors including embedded RISC architectures, VLIW machines, DSPs and RISC workstations was used to measure the runtimes and code sizes of the benchmarks (see section 3.3.2 for a detailed list of processors). Since a detailed profiling including pipeline and cache statistics can not be performed for most of the considered processors due to lacking hardware support, only the runtimes of the benchmarks are reported in the following.

As can be seen in figure 5.15, the runtimes of the CAVITY benchmark are improved between 7.7% (TI C6x) and 35.7% (HP). On average over all processors, a speed-up of 25.1% was measured. The fact that loop nest splitting is able to generate a very regular control flow in the innermost loop of the ME benchmark leads to very high gains for this benchmark. This application is accelerated by 63.2% on average. The minimum speed-up amounts to 36.5% (TriMedia), whereas the Sun platform honors the optimization with an acceleration of 75.8%. For the QSDPCM benchmark, the improvements range from 3% (Power-PC) up to 63.4% (MIPS). On average, loop nest splitting leads to an acceleration of 29.1% for QSDPCM.

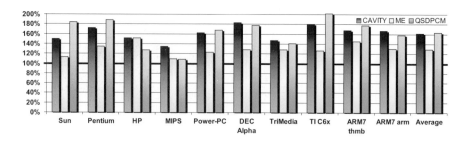

Figure 5.16.   Relative Code Sizes after Loop Nest Splitting

As for every code replicating optimization, the large improvements of loop nest splitting with respect to execution times entail increases in code sizes. In the case of the CAVITY benchmark, the average speed-up of 25.1% comes along with an average increase of code sizes of 61.7% (see figure 5.16). The assembly code generated by the MIPS compiler after loop nest splitting is the most condensed one, since it is only 34.7% larger than the original code. In the case of the DEC Alpha, an increase of 82.8% of code size was measured. Among all benchmarks considered in this section, it is the ME application for which loop nest splitting leads to the highest accelerations. In addition, figure 5.16 shows that the optimization of this benchmark also leads to the smallest code size increases. On average, an increase of only 29.1% was observed whereby the minimum and maximum growths amount to 9.2% (MIPS) and 51.4% (HP). Finally, the increases of code size for the QSDPCM application vary between 8.7% (MIPS) and 101.6% (TI C6x) leading to an average augmentation of 63.2%.

It should be kept in mind that the benchmarks considered here represent the most computation intensive kernels of larger data dominated embedded applications. When considering the entire CAVITY application, the average code size increase only amounts to 36.7%. For a complete MPEG 4 encoder, loop nest splitting of the motion estimation kernel increases code size by only 2.7%. The growth of the QSDPCM application after loop nest splitting only amounts to 5.2%.

However, if code size increases are critical, it is easy to change the genetic algorithms so that the splitting-if is not placed in the outermost possible loop. This way, code duplication is reduced at the expense of lower speed-ups so that trade-offs between code sizes and savings in runtimes can be realized. The analysis of such trade-offs is part of the future work on loop nest splitting.

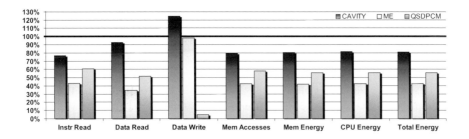

Figure 5.17. Relative Energy Consumption after Loop Nest Splitting

5.5.1.3 Energy Consumption

The increases of code sizes by a few hundred instructions (see previous section 5.5.1.2) are not a serious drawback of loop nest splitting, since the added energy required for storing these instructions is compensated by the savings achieved during the execution of the optimized benchmarks. The results presented in this section demonstrate that loop nest splitting is a powerful technique for reducing the energy dissipation of embedded multimedia applications. Using the instruction-level energy model presented by [Steinke et al., 2001a] for an ARM7TDMI embedded RISC core, the energy consumption of the benchmarks before and after loop nest splitting can be computed with an accuracy of 1.7%.

Using the design flow presented in section 3.3.3, statistical data about the number of executed instructions, memory accesses and about the energy consumption of memories and the CPU can be obtained, which is illustrated in figure 5.17. As in the previous sections, the bars of this figure denote the behavior of the benchmarks after loop nest splitting given as a percentage relative to the unoptimized benchmarks which are represented by 100%.

The first four columns of figure 5.17 confirm the observations already presented in section 5.5.1.1 that loop nest splitting leads to a higher locality of instructions and data thereby reducing memory bandwidth and bus demand. In the case of the ARM7 processor, between 23.5% (CAVITY) and 56.9% (ME) fewer instructions are fetched from memory. Accesses to data stored in the memory are likewise optimized. From column Data Read, it can be seen that data transfers from the memory to the ARM7 core are reduced between 7.1% (CAVITY) and 65.3% (ME). Since data is read much more frequently than it is written to memory, the fact that 24.5% more memory stores occur in the case of CAVITY is not significant. This increase is due to the fact that the used compiler needs to insert some spill code during register allocation. The contrary holds for the QSDPCM benchmark, where the elimination of spill code leads to a reduction of memory stores by 95.4%. When focusing on the total energy dissipation of the entire memory system, only the absolute number of all memory accesses is important. Column Mem Accesses clearly shows that loop nest splitting leads to large reductions, ranging from 20.8% (CAVITY) up to 57.2% (ME).

These improvements consequently lead to large diminutions of the energy dissipation of the memory (see column Mem Energy). The energy profiling shows that loop nest splitting is able to achieve savings between 19.6% (CAVITY) and 57.7% (ME). In addition, the benefits of the transformation are not limited to the memory. As can be seen from column CPU Energy, the amount of energy consumed by the ARM7 core itself is reduced by the same order of magnitude; gains reaching from 18.4% (CAVITY) up to 57.4% (ME) were observed. Accumulated, the total energy consumption of the processor and its memory is reduced between 19.2% (CAVITY) and 57.6% (ME). Among other things, these results also demonstrate that loop nest splitting is capable of optimizing the locality of instruction and data accesses simultaneously.

5.5.2   Combined Data Partitioning and Loop Nest Splitting for energy-efficient Scratchpad Utilization

In order to demonstrate that loop nest splitting is a powerful optimization technique capable of removing negative side-effects from typical embedded software which are caused by data layout transformations, experimental results for the combined application of data partitioning and loop nest splitting are presented in this section. In the following, a system architecture based on an ARM7TDMI processor is assumed. As motivated in section 5.1.2, a typical memory hierarchy consisting of a background main memory and a small on-chip scratchpad memory completes the architecture.

For achieving an optimal scratchpad utilization with minimized energy dissipation, a set of four different representative benchmarks is first processed by the algorithms for data partitioning originally presented in [Verma et al., 2003]. Using integer linear programming techniques, it is decided which arrays of the benchmarks have to be split at which locations so that fragments of the originally large arrays can be accessed via the scratchpad memory with low costs in terms of energy. For a given partitioning decision, the source codes of the benchmarks are transformed as illustrated in figure 5.2 (see page 57) in order to reflect the data partitioning.

In the second step, these partitioned source codes are fed into the tool for loop nest splitting in order to minimize the execution of the *if*-statements generated during data partitioning. All intermediate codes of the benchmarks (i. e. the unchanged code, the code after data partitioning and after loop nest splitting) are compiled using the energy-aware research compiler *encc* so that the efficiency of the benchmarks can be evaluated (compare section 3.3.3).

Since data partitioning and loop nest splitting are performed at the level of C source codes in this book, the entire optimization framework can easily be ported to other system architectures by simply providing an adapted energy model.

For the experiments described in the following, benchmarks from different domains were selected. First, a 40 order *FIR* filter is used as an example for a typical embedded DSP algorithm. Second, the sorting algorithms insertion sort (*INS*) and selection sort (*SELS*) were analyzed. Finally, a complete MPEG4 motion estimation routine (*ME*) was studied.

The relevance of the extensions of loop nest splitting supporting non-constant loop bounds (see section 5.4) is clearly demonstrated by the fact that a splitting of the FIR, INS and SELS benchmarks is impossible using only the techniques requiring constant loop bounds. Although the genetic algorithm for condition optimization supporting non-constant loop bounds has a higher complexity than the one employed for loop nests with constant bounds, the runtimes for applying loop nest splitting to the benchmarks presented above are still very low – not more than 0.17 CPU seconds are required for the execution of the loop nest splitting algorithms on an Intel Pentium IV running at 2.6 GHz.

The experimental results presented in the following sections base on a fixed scratchpad size for each individual benchmark. A memory size of 1.8 kB was used for FIR, and 1.3 kB were used for INS and SELS. The ME routine with its large video frames was analyzed using a memory size of 119 kB. In order to demonstrate the stability of the combined optimization approach consisting of data partitioning and loop nest splitting, results for a large variety of scratchpad sizes using the SELS benchmark are also provided in the following.

In the following sections 5.5.2.1 and 5.5.2.2, the execution times, code sizes and energy consumption of the benchmarks after data partitioning and loop nest splitting are presented.

5.5.2.1   Execution Times and Code Sizes

With respect to the runtimes of the benchmarks (see figure 5.18), the combination of data partitioning and loop nest splitting is quite beneficial. Due to the fact that *if*-statements are inserted in the code of the benchmarks during data partitioning (compare figure 5.2), the execution times of almost all benchmarks increase by 8.7% (SELS) up to 36.7% (INS). Only in the case of the data-intensive ME benchmark, a speed-up of 28.2% was measured. This is due to the fact that ME is very data-intensive and needs to access memory frequently. Since the ARM7 processor accesses an on-chip memory much faster than the main memory, data partitioning in this case has the effect that the number of time consuming wait states executed by the processor is reduced. Additionally, the compiler inserts less spill code after data partitioning leading to the measured runtime improvement.

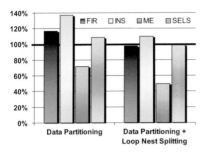
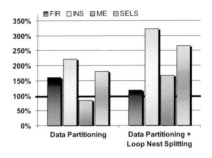

Figure 5.18.   Relative Runtimes after Data Partitioning and Loop Nest Splitting

Figure 5.19.   Relative Code Sizes after Data Partitioning and Loop Nest Splitting

Loop nest splitting is able to eliminate the negative effects of data partition-ing nearly totally. Compared to the runtimes after data partitioning, speed-ups between 9.3% (SELS) and 31.3% (ME) were measured. When considering the entire flow of optimization, it can be observed that the runtimes of two bench-marks are improved slightly (FIR: 2.5%, SELS: 1.5%) after the application of data partitioning and loop nest splitting. For the INS benchmark, a moderate total runtime degradation of 9.7% was still measured after loop nest splitting. In this single case, loop nest splitting is only able to eliminate the overhead in-troduced by data partitioning to a large extent, but not completely. In contrast, the ME benchmark was accelerated by 50.6% overall.

In analogy to the results already presented in section 5.5.1.2, increases of code sizes can be observed after the application of both optimization techniques as illustrated in figure 5.19. It can be seen that data partitioning leads to growths of code sizes between 60.9% (FIR) and up to 121.7% (INS). For the ME bench-mark, a decrease of code size of 15.5% was measured which is caused by the insertion of less code for register spilling after data partitioning. Loop nest splitting applied after data partitioning generally leads to additional code size increases. After the entire optimization process, the generated assembly codes are between 17.4% (FIR) and 221.8% (INS) larger than the original unoptimized code versions. On the average over all four benchmarks, data partitioning leads to a code size increase of 61.8%, whereas loop nest splitting only contributes by 55.9% on average. These code size increases imposed by loop nest splitting are of the same order of magnitude as already presented in section 5.5.1.2.

Since in the worst case, loop nest splitting might lead to the duplication of an entire loop nest in the *then-* and *else-*part of the splitting-if except the outermost loop, a rough theoretical upper bound of code size increase after loop nest splitting is 100%. The increases measured and presented in this chapter show that this upper bound is reached only in very few cases in practice

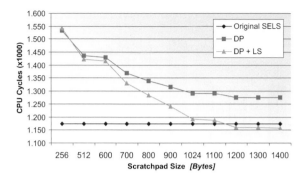

Figure 5.20. Runtime Comparison for varying Scratchpad Sizes (SELS Benchmark)

(QSDPCM on TI C6x in figure 5.16, and INS in figure 5.19). With 53.4% of average code size growth for all benchmarks on all architectures, loop nest splitting stays far below this theoretical maximum.

The influence of varying scratchpad sizes on the execution times of the SELS benchmark is depicted in figure 5.20. This diagram illustrates the total runtimes of this benchmark for eleven different scratchpad sizes. As can be seen from the figure, the original unoptimized code version requires constant execution times for all scratchpad sizes. This is due to the fact that the scratchpad is not used at all for this unpartitioned code so that no accelerations due to the fast memory can be achieved. Starting from significantly increased runtimes for small scratchpads, the overhead of data partitioning gets smaller the larger the on-chip memory becomes. This behavior is due to the high latencies imposed by main memory accesses which are minimized most effectively by data partitioning for larger scratchpads. In contrast, the benefits of loop nest splitting are already visible for scratchpad sizes from 512 bytes. From this point on, the impact of loop nest splitting on the performance of the benchmark becomes more and more significant as the speed-ups increase. For a scratchpad size of 1,024 bytes, a runtime nearly equal to the one of the original code has been obtained. For 1,200 bytes, the code generated by the proposed combined optimization methodology is faster than the original code so that loop nest splitting is able to over-compensate the overheads of data partitioning.

5.5.2.2 Energy Consumption

Figures 5.21 and 5.22 show the effects of the combination of data partitioning and loop nest splitting on the energy consumption of the benchmarks. All results are shown as a percentage of the original unoptimized benchmark codes denoted as 100%. For both data partitioning and loop nest splitting, the relative energy consumed by the memory system (i. e. main memory and on-chip scratchpad),

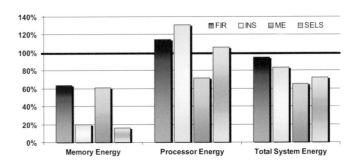

Figure 5.21.   Relative Energy Consumption after Data Partitioning

by the ARM7 processor and by the total system (i. e. processor plus memories) is depicted.

Diagram 5.21 clearly shows that data partitioning is a highly effective optimization regarding the memory. Column Memory Energy shows that the partitioning of arrays and the placement of parts of arrays onto a scratchpad leads to energy savings between 36.7% (FIR) and 84.2% (SELS). Due to the impaired control flow after data partitioning, the energy consumed by the ARM7 processor generally increases when compared to the original code version. Column Processor Energy shows increases between 6% (SELS) and 30.8% (INS). In the case of the ME benchmark, an energy reduction of 28.6% was measured. As already explained in the previous section, this behavior is due to less wait states and register spills after data partitioning. For the entire system (column Total System Energy), the data partitioning techniques presented in [Verma et al., 2003] lead to total energy savings between 5.7% (FIR) and 34.7% (ME) with an average improvement of 21.3%.

Figure 5.22 illustrates the relative energy consumption of the benchmarks after combined data partitioning and loop nest splitting. As can be seen by comparing the columns Memory Energy of figures 5.21 and 5.22, loop nest splitting conserves the energy savings for the memory system achieved by data partitioning. In the case of the FIR benchmark, additional savings of memory energy by 9.9% were measured. The significantly reduced *if*-statement executions for this benchmark imply less instruction fetches from the memories leading to this result. Column Processor Energy of figure 5.22 clearly shows that loop nest splitting is able to eliminate the penalties introduced by data partitioning completely. After this source code optimization, the energy consumption of the ARM7 for the FIR and SELS benchmarks even drops below the level of the original unoptimized code. In the case of the INS benchmark, the ARM7 consumes only 5.8% more energy than before any optimization. When comparing the columns Processor Energy of figures 5.21 and 5.22 for the INS benchmark,

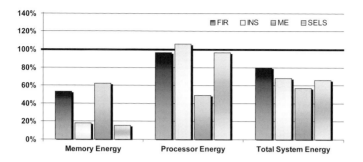

Figure 5.22. Relative Energy Consumption after Data Partitioning and Loop Nest Splitting

a reduction of energy dissipation by 25% can be observed after loop nest splitting. For the ME benchmark, loop nest splitting leads to an energy reduction for the ARM7 processor by 31.2% compared to the measurements immediately after data partitioning. Column Total System Energy illustrates the total savings achieved by the overall flow of optimization. It can be seen that the combined energy dissipation of the ARM7 and its memories drops between 20.3% (FIR) and 43.3% (ME) with an average saving of 32.3% compared to the unoptimized benchmarks.

The results presented in this section demonstrate that a combined optimization methodology consisting of data layout transformations followed by control flow improvement is highly beneficial. Data partitioning leads to average reductions of energy dissipation of 21.2% for the ARM7 based platform considered here. Loop nest splitting is able to remove the control flow overheads resulting from data partitioning so that the runtimes are improved significantly for all benchmarks. Furthermore, the techniques presented in this chapter achieve an additional reduction of energy consumption of 13% on average. Data partitioning and loop nest splitting complement each other in such a way that both lead to considerable energy savings, while the combined application of the optimizations ensures that the finally generated code contains only a minimum of control flow overhead, but a maximum of data stored in low energy scratchpad memories.

Figure 5.23 illustrates the impact of different scratchpad sizes on the energy consumption of the SELS benchmark. It is not surprising that the original unoptimized benchmark consumes the same high amount of energy for all scratchpad sizes. This is due to the fact that no data can be placed on the scratchpad memory at all due to the large size of the occurring arrays. In contrast, data partitioning is effective in energy consumption minimization already for very small memory sizes. In the case of a 256 bytes memory, only negligible improvements were measured which are not visible due to the resolution of figure 5.23. But

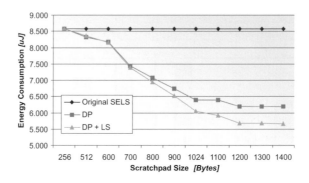

Figure 5.23. Energy Comparison for varying Scratchpad Sizes (SELS Benchmark)

already for 512 bytes, significant improvements were observed. With larger
scratchpad sizes, data partitioning achieves higher gains due to the fact that less
costly accesses to the main memory are performed. This way, a monotonically
decreasing curve has been obtained. The same holds for loop nest splitting
applied after data partitioning. Here, loop nest splitting is able to reduce the
energy consumption considerably for scratchpads larger than 600 bytes. Again,
a monotonic regression can be observed clearly demonstrating the stability of
the combined optimization methodology.

5.6   Summary

In this chapter, a novel source code optimization for control flow improvement of data flow dominated embedded software called loop nest splitting was
presented. It bases on the observation that *if*-statements occur frequently in
deeply nested loops which check conditions depending on the index variables
of the loop nest. Loop nest splitting is able to detect ranges of iterations of a
loop nest where these *if*-statements are provably satisfied. Using these ranges
of iterations, a loop nest is optimized in such a way that after the transformation,
the number of executed *if*-statements is minimized.

The benefits of loop nest splitting were proven to be manifold. A very detailed
benchmarking showed that a massive reduction of executed branch instructions
and of pipeline stalls is caused by the minimized execution of *if*-statements.
Since the execution of *if*-statements is minimized, the execution of all the code
related to the evaluation of the conditions of these *if*-statements is automatically
minimized leading to further performance improvements. In addition, memory
accesses were reduced by loop nest splitting, because less transfers of variables
and instructions between processor and memory have to be performed. In total,
all these factors lead to significant speed-ups and energy savings achieved by
loop nest splitting.

The optimization and analysis techniques presented in this chapter are kept very general so that they are able to optimize large classes of loop nests. On the one hand, these techniques can be applied to general data dominated software which often contains *if*-statements in loop nests (e. g. MPEG4 motion estimation). On the other hand, source codes generated by data layout transformation frameworks (e. g. DTSE [Catthoor et al., 1998], data partitioning [Verma et al., 2003]) can successfully be optimized using loop nest splitting.

The techniques presented in this chapter were published in [Falk and Verma, 2004], [Falk et al., 2003b], [Falk and Marwedel, 2003] and [Falk, 2002].

Chapter 6

ADVANCED CODE HOISTING

In this chapter, a source code optimization called *advanced code hoisting* is presented. This technique aims at moving portions of code from inner loops to outer ones. In contrast to existing code motion techniques, this is performed under consideration of control flow aspects. Depending on the conditions of *if*-statements, moving an expression can lead to an increased number of executions of this expression. This chapter contains formal descriptions of the models used for control flow analysis so as to suppress a code motion in such a situation.

In order to provide an introduction into the concepts of advanced code hoisting, this transformation is presented in section 6.1 by means of a simple example. Since the code structures explored by advanced code hoisting are generated by other source code transformation frameworks, an overview of related work is given in section 6.2. The models and algorithms employed for code hoisting under consideration of the control flow are presented in detail in section 6.3. In section 6.4, experimental results are provided, followed by a summary in section 6.5.

6.1   A motivating Example

Figure 6.1 shows a source code fragment extracted from the innovative GSM codebook search [3GPP, 1999]. In this code, array elements are assigned to a temporary variable rdm. The selection of these array elements depends on the index variables of the surrounding *for*-loops. Additionally, the assignment of values to the variable rdm is controlled by nested *if*-statements.

As can be seen from this source code, the expressions used to address the array rr are complex and contain costly arithmetical operations like integer multiplication and division. The goal of the advanced code hoisting optimization is to minimize this large overhead caused by address computations. This is done by moving code from inner loops to outer ones as long as this leads to

```
for (j=0; j<l_code; j++)
  for (i=0; i<l_code; i++)
  {
    if (i<=c1)
      if (i==c1)
        rdm = h2[l_code-1-i];
      else rdm = rr[i*(i-1)/2];
    else rdm = rr[j*(j-1)/2];
    ...mul(rdm, ...);

    if (i<=c2)
      if (i==c2)
        rdm = h2[l_code-1-i];
      else rdm = rr[j*(j-1)/2];
    else rdm = rr[i*(i-1)/2];
    ...mul(rdm, ...);
  }
```

Figure 6.1.   A Code Sequence of GSM Codebook Search

a reduced number of executions of the code. More precisely, *common subexpressions* are considered as candidates for advanced code hoisting:

DEFINITION 6.1 (COMMON SUBEXPRESSION)
*An occurrence of an expression in a program is a **common subexpression** if there exists another occurrence of the same expression whose evaluation always precedes this one in execution order and if the operands of the expression remain unchanged between the two evaluations.*

Advanced code hoisting performs its optimization in a three-step approach:

1 Common subexpressions present in a source code are detected.

2 For every subexpression determined in the first step, the outermost loop which can legally contain it is identified.

3 A new variable is defined in the loop computed in step 2, and the common subexpression is assigned to it. Every occurrence of the subexpression in the program is replaced by this new variable. This step is performed only when it leads to a total reduction of computations of the common subexpression.

For the example shown in figure 6.1, the first step leads to the identification of the expressions

- l_code-1-i,

- `i*(i-1)/2` and

- `j*(j-1)/2`.

In the second step, a data flow analysis applied to these expressions leads to the result that

- `l_code-1-i` can be moved to the beginning of the `i`-loop,

- `i*(i-1)/2` can be moved to the beginning of the `i`-loop and

- `j*(j-1)/2` can be moved to the beginning of the `j`-loop.

The final decision whether a common subexpression is hoisted to the positions determined in step 2 depends on the number of executions of the expression before and after the code motion. At this stage of the optimization process, a control flow analysis is performed considering *for*-loops and *if*-statements surrounding the common subexpressions. Assuming that the symbolic constants `l_code`, `c1` and `c2` are equal to 10,000, 9,000 and 1,000 respectively, the code shown in figure 6.1 leads to the following number of executions:

- `l_code-1-i` is executed 20,000 times,

- `i*(i-1)/2` is executed 179,990,000 times and

- `j*(j-1)/2` is executed 19,990,000 times.

If all expressions listed above are moved to the aforementioned positions, the source code depicted in figure 6.2 will result. For the given values of `l_code`, `c1` and `c2`, this source code leads to the following execution frequencies of the identified common subexpressions:

- `l_code-1-i` is computed 100,000,000 times,

- `i*(i-1)/2` is computed 100,000,000 times and

- `j*(j-1)/2` is computed 10,000 times.

The key idea of advanced code hoisting is to compare the execution frequencies of the common subexpressions before and after their motion to the outermost loops without consideration of control flow issues. For the given example, it can be observed that `i*(i-1)/2` and `j*(j-1)/2` are executed less frequently after performing code motion. In contrast, the code hoisting of `l_code-1-i` is not beneficial since it leads to a large increase in the number of computations of this expression. As a consequence, advanced code hoisting takes the decision to move `i*(i-1)/2` and `j*(j-1)/2` to the positions shown in figure 6.2, whereas the expression `l_code-1-i` is not moved at all. The resulting source code is illustrated in figure 6.3.

```
for (j=0; j<l_code; j++)
{
  int tmp1 = j*(j-1)/2;

  for (i=0; i<l_code; i++)
  {
    int tmp2 = l_code-1-i, tmp3 = i*(i-1)/2;

    if (i<=c1)
      if (i==c1)
        rdm = h2[tmp2];
      else rdm = rr[tmp3];
    else rdm = rr[tmp1];
    ...mul(rdm, ...);

    if (i<=c2)
      if (i==c2)
        rdm = h2[tmp2];
      else rdm = rr[tmp1];
    else rdm = rr[tmp3];
    ...mul(rdm, ...);
  }
}
```

Figure 6.2.   The GSM Codebook Search after Code Hoisting without Control Flow Analysis

The reason for the increased execution frequency of $l\_code-1-i$ after its motion to the beginning of the i-loop are the two *if*-statements surrounding the common subexpression. In the original code depicted in figure 6.1, the code lines rdm = h2[$l\_code-1-i$] are executed only if the index variable i is equal to $c1$ and $c2$ respectively. Trivially, these conditions are true for only two iterations of the entire i-loop. Since the i-loop is nested in the j-loop, a total of 20,000 executions of the expression $l\_code-1-i$ is the result.

After moving the expression $l\_code-1-i$ to the beginning of the i-loop, this expression is no longer guarded by these two *if*-statements. Thus, the expression is computed during every iteration of the i-loop resulting in the observed execution frequency of 100,000,000.

The influence of *if*-statements surrounding common subexpressions can be illustrated further when assuming other values of $c1$ and $c2$. For $c1$ equal to 5,001 and $c2$ equal to 4,999, $i*(i-1)/2$ and $j*(j-1)/2$ are executed

```
for (j=0; j<l_code; j++)
{
    int ach_1 = j*(j-1)/2;

    for (i=0; i<l_code; i++)
    {
        int ach_2 = i*(i-1)/2;

        if (i<=c1)
            if (i==c1)
                rdm = h2[l_code-1-i];
            else rdm = rr[ach_2];
        else rdm = rr[ach_1];
        ...mul(rdm, ...);

        if (i<=c2)
            if (i==c2)
                rdm = h2[l_code-1-i];
            else rdm = rr[ach_1];
        else rdm = rr[ach_2];
        ...mul(rdm, ...);
    }
}
```

Figure 6.3.   The GSM Codebook Search after Advanced Code Hoisting

100,010,000 and 99,970,000 times respectively for the original source code[1].
After hoisting these expressions to the beginning of the j- and i-loops, execution frequencies of 100,000,000 and 10,000 are observed. For this example, a
reduction of executions of the two expressions can still be observed so that the
code motion is performed. However, these reductions are less than the ones
reported for the previous example due to the changed values of $c1$ and $c2$.

The situation changes for $c1$ and $c2$ both having the value of 5,000. For these
parameters it can be seen that the two expressions i*(i-1)/2 and j*(j-1)/2
are both executed 99,990,000 times. After their motion to the beginning of the
loops, these expressions are executed 100,000,000 and 10,000 times again. For
i*(i-1)/2, this is an increase of executions so that advanced code hoisting is
only performed for the expression j*(j-1)/2.

The first main contribution of the advanced code hoisting techniques presented in this chapter is the novel combination of well-known compiler opti-

---

[1] l_code-1-i is not considered here since its execution frequencies of 20,000 and 100,000,000 remain
unchanged.

```
for (y=0; y<=M; y++)                for (y=0; y<=M; y++)
   for (x=0; x<=N; x++) {              for (x=0; x<=N; x++) {
      if (x>=2 && y<=M-2)                 if (x>=2 && y<=M-2)
         gxc += pixels[(x-2)%3];             gxc += pixels[(x-2)%3];
      if (y>=1 && x<N+1) {    ⟶          if (y>=1 && x<N+1) {
                                              int cse1 = (y%2)*N;

      ... = a[(x-2)%3+(y%2)*N];            ... = a[(x-2)%3+cse1];
      a[(y%2)*N] += ...; }                 a[cse1] += ...; }
   }                                  }
```

Figure 6.4.   Local Common Subexpression Elimination

mizations – namely common subexpression elimination and loop-invariant code motion – with a criterion steering the application of these two optimizations. Second, the formulation of this criterion deciding when to apply code motion is based on control flow aspects. In this chapter, formal methods are described in order to model large classes of loop nests and *if*-statements. Polyhedral techniques are applied for the exact computation of execution frequencies of expressions. The experimental results provided in this chapter show that advanced code hoisting is able to achieve large improvements with respect to execution times and energy consumption. Third, all compilers involved during benchmarking are called with their highest levels of optimization enabled. Since they all include a common subexpression elimination and a loop-invariant code motion, it is an important observation that still very large savings can be achieved using advanced code hoisting. This fact underlines that novel source code optimizations which rely on already existing techniques to some extent are still able to outperform existing optimizing compilers.

6.2   Related Work

Since common subexpressions and loop-invariant code play a major role in the context of advanced code hoisting, two well-known compiler optimizations also focusing on these code structures are presented first [Bacon et al., 1994, Muchnick, 1997].

Common subexpression elimination is a transformation focusing on redundancy elimination. According to definition 6.1, all occurrences of a common subexpression (*CSE*) necessarily compute the same result since all operands of the expression remain unchanged during the execution of the program code lying in-between these CSE occurrences. The repeated recomputation of the value of a subexpression is redundant since it is sufficient to compute the value of a CSE once, store it in a temporary variable and reuse the stored result. This

```
for (y=0; y<=M; y++)              for (y=0; y<=M; y++)
   for (x=0; x<=N; x++) {            for (x=0; x<=N; x++) {
                                        int cse2 = (x-2)%3;

      if (x>=2 && y<=M-2)              if (x>=2 && y<=M-2)
         gxc += pixels[(x-2)%3];          gxc += pixels[cse2];
      if (y>=1 && x<N+1) {     →        if (y>=1 && x<N+1) {
                                           int cse1 = (y%2)*N;

         ... = a[(x-2)%3+(y%2)*N];        ... = a[cse2+cse1];
         a[(y%2)*N] += ...; }             a[cse1] += ...; }
   }                                 }
```

Figure 6.5. Global Common Subexpression Elimination

process of storing intermediate results in temporary variables is called elimination of common subexpressions. If storing temporary values in variables forces additional spills to memory due to a high register pressure, the transformation can actually degrade a program's performance.

In the area of compiler design, common subexpression elimination is frequently divided into two phases [Muchnick, 1997]: during *local common subexpression elimination*, the search for CSEs and their elimination is performed only at the level of basic blocks (see definition 4.1 on page 45). Since basic blocks are maximal code sequences with linear flow of control, it is easy to decide whether two expressions in a basic block precede each other and whether the instructions in-between modify an operand of the expression. As a consequence, a local common subexpression elimination can be done with little effort while the intermediate code for a basic block is being constructed.

In contrast, a *global common subexpression elimination* has to consider modifications of the control flow between two occurrences of an expression making it more complicated to determine the execution order of two expressions. On the other hand, a global common subexpression elimination is able to catch all the CSEs that the local form does and more, so that the global approach is more powerful.

The differences between local and global common subexpression elimination are depicted in figures 6.4 and 6.5 by means of a simple source code fragment. Due to the fact that the local form is unable to cross control flow boundaries, it only recognizes and eliminates the expression (y%2)*N, because all occurrences of this expression are within the same *if*-statement (see figure 6.4). The expression (x-2)%3 can only be eliminated by a global optimization since both occurrences of this expression reside in different *if*-statements. Additionally, the global technique also manipulates the expression (y%2)*N as the local com-

```
                                    int t1 = 10*(n+2), t2 = 100*n;
for (i=1; i<=100; i++) {
                                    for (i=1; i<=100; i++) {
    l = i*(n+2);
                                        int t3 = t2+i*t1;
    for (j=i; j<=100; j++)
                              →
        a[i,j] = 100*n+10*l+j;          for (j=i; j<=100; j++)
}                                           a[i,j] = t3+j;

                                    }
```

Figure 6.6.   Loop-invariant Code Motion

mon subexpression elimination does. The resulting source code is depicted on
the right hand side of figure 6.5.

Loop-invariant code motion recognizes computations in loops that produce
the same value on every iteration of the loop and moves them out of the loop [Aho
et al., 1986, Bacon et al., 1994, Muchnick, 1997]. If a computation occurs inside
a nested loop, it may produce the same value for every iteration of the inner
loops for each particular iteration of the outer loops, but different values for
different iterations of the outer loops. Such a computation will then be moved
out of the inner loops, but not the outer ones. This process is illustrated in
figure 6.6.

As can be seen easily, the value of the expression 100*n+10*l does not
change during any iteration of the j-loop. Since this expression depends on
the variable l which is recomputed during every iteration of the i-loop, these
computations can be moved out of the j-loop. Hereafter, the loop-invariant
expressions 10*(n+2) and 100*n are moved out of the entire loop nest. The
right hand side of the above figure shows the resulting code.

The identification of loop-invariant code can be performed easily. An expression is loop-invariant if, for each of its operands [Muchnick, 1997]:

- the operand is constant,

- all definitions of the operand are located outside the loop, or

- there is exactly one definition of the operand, and this definition is located
  in the same loop as the entire expression and is itself loop-invariant.

It should be emphasized at this place that both common subexpression elimination and loop-invariant code motion are especially beneficial when applied
to addressing code as stated explicitly in [Bacon et al., 1994, Muchnick, 1997].
Since addressing code is generally not written explicitly by human programmers
but is generated automatically by a compiler, the programmer is often unaware
of the overhead due to memory accesses. In the remainder of this section,
parts of the already mentioned DTSE source code transformation framework

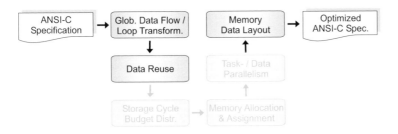

Figure 6.7. Parts of DTSE causing Control Flow and Addressing Overhead

are presented in more detail. Since DTSE aims at the optimized exploitation of memory hierarchies, the DTSE transformations have the effect of making addressing code explicit in a program. In this context, techniques for address code optimization were proposed in the ADOPT project [Miranda et al., 1998]. In their combination, DTSE and ADOPT motivate the need for further optimization of address code at the source code level. Advanced code hoisting explicitly focuses on the optimization of DTSE-generated code.

Figure 6.7 highlights those parts of the DTSE methodology that are responsible for control flow and addressing overheads which are the target of advanced code hoisting as motivated in section 6.1. During the stage of *global data flow and loop transformation*, an "advanced signal substitution" is performed. The goal of this technique is to remove dependencies on multi-dimensional signals and to eliminate redundant array accesses. In [Catthoor et al., 1998], the example of an H.263 video decoding software is given where the data stored in borders of images is duplicated and thus stored in an additional "edge buffer". Signal substitution is able to reduce this overhead by analyzing the data dependencies between the pixels stored in the original video frame and in the additional edge buffer. By inserting *if*-statements depending on the loop index variables, these data dependencies can be resolved so that always the original pixels of the frames' edges can be accessed. This way, all accesses to the edge buffer can be removed leading to the complete removal of this array.

During *data reuse exploration*, frequently accessed arrays are analyzed in order to exploit the temporal locality of array accesses under consideration of memory hierarchies. The main idea of this technique is to introduce copies of the most frequently accessed parts of arrays, i. e. to establish a "copy-candidate chain" [Wuytack et al., 1998]. The size of these copies is determined such that every copy fits into a smaller and thus faster and less energy consuming memory. All accesses to the identified parts of the original array are replaced by accesses to the smaller copy. Data reuse exploration is a powerful technique which is able to keep non-sequential parts of an original array in a copy. This technique introduces code overhead since all copies need to be filled with meaningful

```
                                    for (y=0; y<M+2; y++)
                                      for (x=0; x<N+2; x++) {
  for (x=1; x<N-2; x++)                 ...
    for (y=1; y<M-2; y++) {             if (x>=0 && x<N && y>0 && y<M-1)
      for (k=-1; k<1; k++) {              D[x%3]=B[96+(y*N+x%3)%160+
        ...                    →                (y*N+x%3)/160*256];
      A[x][y]+=B[x+k][y];                ...
      ... }                             if (x>1 && x<N && y>0 && y<=M-2)
    A[x][y]/=tot; }                       for (k=-1; k<=1; k++)
                                            acc+=D[(x-1+k)%3];
                                      acc/=tot; }
```

Figure 6.8.   A Fragment of the CAVITY Benchmark before and after DTSE

content which has to be written back to the original array after manipulation. In addition, *if*-statements depending on the loop iterations are potentially inserted again so as to select all relevant non-contiguous parts that will be held in a copy.

The "in-place mapping" stage is part of the *memory data layout optimizations* shown in figure 6.7. This technique aims at reusing physical memory locations as much as possible so that several data entities are stored at the same locations, but at different times [Greef et al., 1997]. Among others, in-place mapping techniques were presented which are applied to individual arrays. In this context, the lifetimes of array elements are analyzed, and elements that are not alive at the same time can be mapped to the same position within the array. This way, array sizes are scaled down and conflict cache misses are reduced. This optimization requires the generation of complex addressing code reflecting the mapping of data elements to positions within an array. The effect of these parts of the DTSE methodology on a source code taken from [Miranda et al., 2001] is illustrated in figure 6.8.

Since many different data elements are stored in the same array addresses after in-place mapping, the same kind of address computations is performed several times at various locations in a DTSE optimized code. This is already acknowledged by work in the context of ADOPT (address optimization). In [Janssen, 2000], a series of algebraic transformations for regularity improvement is presented. During regularity improvement, a given set of address expressions $A = \{a_1, \ldots, a_n\}$ is transformed to $A' = \{a'_1, \ldots, a'_n\}$. Every expression a'_i computes the same value as a_i, but A' is generated such that a maximal reuse of address computations using CSEs is achieved:

```
  ... = B[(y-1) % 3];          ... = B[(y-1) % 3];
  A[(y-4) % 3] = ...;    →     A[(y-1) % 3] = ...;
```

```
advanced_code_hoisting() {
  for (all functions F) {
    expression_list CSE_list = identify_CSEs(F);

    for (all CSEs CSE ∈ CSE_list) {
      position outermost_position;
      int execs_before = 0, execs_after;

      for (all occurrences of CSE CSE)
        execs_before += compute_executions(occurrence of CSE);

      outermost_position = hoisted_position(CSE);
      execs_after = compute_executions(outermost_position);

      if (execs_after < execs_before)
        eliminate_CSE(CSE, outermost_position);
    }
  }
}
```

Figure 6.9. Algorithm for Advanced Code Hoisting

This is done by applying a series of simple elementary transformations in an optimized manner to the address expressions:

$$(y-4) \% 3 \quad \rightarrow \quad (y\%3-4\%3) \% 3 \quad \rightarrow \quad (y\%3-1) \% 3$$
$$\rightarrow \quad (y\%3-1\%3) \% 3 \quad \rightarrow \quad (y-1) \% 3$$

These algebraic transformations open opportunities for common subexpression elimination as stated in [Gupta et al., 2000]. For the ADOPT transformations, experiments involving the manual application of a global common subexpression elimination combined with conventional loop-invariant code motion are reported. It is the contribution of this chapter that a formal problem definition for advanced code hoisting is provided for the first time in conjunction with suitable algorithms considering control flow issues.

6.3 Analysis Techniques for Advanced Code Hoisting

In this section, all analysis and optimization techniques required for advanced code hoisting are presented. As mentioned previously, this optimization is performed in three sequential steps:

1 Common subexpressions present in a source code are identified.

2 For every CSE, the outermost loop which is able to store the CSE is determined.

```
expression_list identify_CSEs(function F) {
  expression_list CSE_list = build_CSE_list(F);

  for (all expressions expr_i ∈ CSE_list)
    compute_live_ranges(expr_i);

  return CSE_list;
}
```

Figure 6.10.   Algorithm to identify Common Subexpressions

3 The execution frequencies of the CSE before and after code hoisting are
   computed. If a CSE is executed less after advanced code hoisting, the
   elimination and hoisting of the CSE is performed.

Illustrated using pseudo-code, the algorithm for advanced code hoisting has
the shape shown in figure 6.9.

Phase 1 of this approach is explained in section 6.3.1. The process of determination of an outermost loop for an identified CSE according to the principles of
advanced code hoisting is described in section 6.3.2. Section 6.3.3 is dedicated
to the computation of execution frequencies using polytope models.

6.3.1   Common Subexpression Identification

The process of identification of common subexpressions according to definition 6.1 is performed by collecting all equivalent expressions and their occurrences in a source code during a first phase (see section 6.3.1.1). In a second
step, all occurrences of an expression are grouped together such that all expression occurrences belonging to the same group always compute the same
result (cf. section 6.3.1.2). All such expression occurrences finally constitute
a common subexpression.

6.3.1.1   Collection of equivalent Expressions

In order to identify common subexpressions globally, the code of an entire
ANSI-C function F present in the SUIF intermediate format (see section 4.3.1
on page 47) is traversed by the algorithm presented in figure 6.10.

In a first step, a list of expressions is generated by build_CSE_list which is
specified in the following definition 6.2. After that, this list is traversed and an
analysis is performed in order to determine those ranges of code within function
F where an expression is alive, i. e. remains unchanged.

DEFINITION 6.2 (LIST OF COMMON SUBEXPRESSIONS)
CSE_list $= (expr_1, \ldots, expr_n)$ *is a list of length n containing all expressions*

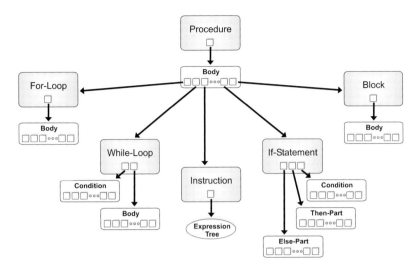

Figure 6.11. The SUIF Abstract Syntax Tree

extracted from the SUIF representation of an ANSI-C function F. For every expression $expr_i \in CSE\_list$, $1 \leq i \leq n$, the following properties are required:

a *$expr_i$ exclusively consists of those parts of the SUIF abstract syntax tree representing instructions. Control flow constructs (e. g. loops or if-statements) must not be present in $expr_i$.*

b *$expr_i$ is only stored once in CSE_list.*

c *$expr_{i,1}, \ldots, expr_{i,\#o_i}$ denotes all occurrences of $expr_i$ in F. The number of occurrences of $expr_i$ in F is denoted by $\#o_i$.*

d *$\#o_i > 1$*

As already mentioned in section 4.3.1, the abstract syntax tree of SUIF contains explicit control flow constructs since SUIF is a high-level intermediate representation (compare figure 6.11). In order to guarantee that only valid computational expressions are considered during CSE identification, only the leaves of the SUIF syntax tree representing pure instructions are considered as stated in definition 6.2a.

Item b of definition 6.2 ensures that no duplicates of the same common subexpression are kept in *CSE_list*. For this purpose, an equivalence check between any new expression *expr* and all $expr_i \in CSE\_list$ has to be performed. For an efficient lookup of positions of $expr_i \in CSE\_list$ in the code, pointers to the occurrences of $expr_i$ are maintained as described in definition 6.2.c. Part d of definition 6.2 makes sure that only expressions with at least two occurrences are stored in *CSE_list*. This is necessary since any expression with only one occurrence can never be a common subexpression.

In the context of advanced code hoisting, only a trivial equivalence check between two expressions based on their expression trees is performed:

DEFINITION 6.3 (EQUIVALENCE OF EXPRESSIONS)
Let $T_1 = (R_1, S_1')$ and $T_2 = (R_2, S_2')$ be the tree representation of expressions *$expr_1$ and $expr_2$. R_1 and R_2 represent the root nodes of the expression trees, and S_1' resp. S_2' are the lists of all subtrees directly connected to the root. $expr_1$ and $expr_2$ are said to be **equivalent** if*

a *the root nodes R_1 and R_2 represent exactly the same instruction or variable, and*

b *all subexpressions represented by the trees contained in S_1' and S_2' are equivalent pairwise.*

During advanced code hoisting, no techniques are applied in order to identify larger classes of common subexpressions. For example, the associativity of various arithmetical operators is not explored having the effect that the expressions a * b and b * a are not considered to be equivalent. This behavior is not a limitation since it has to be put in the context in which advanced code hoisting is performed. As mentioned previously in section 6.2, advanced code hoisting is especially intended to remove the addressing overhead generated during the DTSE transformations. The aforementioned algebraic transformations for regularity improvement are successful in generating a very uniform shape of expressions. As a consequence, the criterion formulated in definition 6.3 is fully adequate for finding all common subexpressions.

Formally, *CSE_list* by itself is not yet a list of common subexpressions after its construction according to definition 6.2. This is due to the fact that up to this moment, no analysis has been performed whether all operands of an expression *expr* $\in$ *CSE_list* remain unchanged between two occurrences of *expr* in function F (compare definition 6.1). This kind of liveliness analysis of operands is performed as a subsequent phase immediately after the construction of *CSE_list* (see algorithm 6.10).

6.3.1.2   Computation of Live Ranges of Expressions

For this purpose, *CSE_list* is traversed and all expressions $expr_i \in$ *CSE_list* are analyzed sequentially. The goal of the analysis performed for an expression $expr_i$ is to determine maximal live ranges for the occurrences of $expr_i$ in an ANSI-C function F:

DEFINITION 6.4 (LIVE RANGES OF EXPRESSIONS)
Let $expr_i \in$ CSE_list be an expression with at least two occurrences $expr_{i,1}$, *..., $expr_{i,\#o_i}$, $\#o_i > 1$.*

A series of consecutive occurrences $expr_{i,j}, \ldots, expr_{i,k}, 1 \leq j < k \leq \#o_i$, is said to lie within the same **live range** *if*

$$\forall(expr_{i,c}, expr_{i,c+1}): \quad j \leq c < k \Rightarrow$$

The operands of $expr_i$ remain unchanged between the c-th and $c+1$-st occurrence of $expr_i$.

More verbosely, definition 6.4 groups together all those occurrences of an expression $expr_i$ which provably compute the same result since all operands remain unchanged. All the occurrences within the same live range can be legally eliminated by storing the result of $expr_i$ in a temporary variable and by replacing the occurrences within the live range with this variable.

For advanced code hoisting to be as powerful as possible, it is necessary to compute maximum live ranges of expression occurrences. This ensures that a minimum of local variables is inserted under simultaneous maximal reuse of these variables after elimination.

EXAMPLE 6.1

The following code fragment serves as an example for the illustration of maximal live ranges. Here, the expression a * b *occurs six times in the entire code fragment. Whenever omitted code is represented by "...", it is assumed that neither* a *nor* b *are referenced.*

```
1   int A[], B[], a, b;
2   A[...(a*b)...] = ...;
3   ... = B[...(a*b)...];
4   i = ...(a*b)...;
5   a = a+1;
6   B[...(a*b)...] = A[...(a*b)...];
7   b = b+a;
8   A[...(a*b)...] = i;
```

According to definition 6.4, both occurrences of a * b *in line 6 belong to the same live range since* a *and* b *remain unchanged in this line. In contrast, the occurrence of line 8 does not belong to this live range since* b *is altered in line 7.*

Actually, the occurrence of line 8 does not belong to any live range since no other occurrence of a * b *exists that can be grouped together with the one of line 8.*

The expressions a * b *found in lines 2 and 3 obviously belong to the same live range. The same holds for lines 3 and 4. As a consequence, the expressions occurring in lines 2, 3 and 4 belong together. They form a maximal live range since it is impossible to expand it to line 6 due to the modification of* a *in line 5.*

According to the two live ranges detected for the code fragment above, the elimination of a * b *would lead to the following code:*

```
1    int A[], B[], a, b;
2    tmp1 = a*b;
3    A[...tmp1...] = ...;
4    ... = B[...tmp1...];
5    i = ...tmp1...;
6    a = a+1;
7    tmp2 = a*b;
8    B[...tmp2...] = A[...tmp2...];
9    b = b+a;
10   A[...(a*b)...] = i;
```

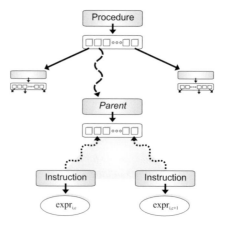

Figure 6.12.   Bottom-up Syntax Tree Traversal

In order to compute maximal live ranges, it is necessary to define exactly which code is being executed between two occurrences of an expression.

A procedure of a high-SUIF syntax tree contains a list storing all subtrees of the procedure body (see figure 6.11). These lists are generically used within the syntax tree. They can consist of all different nodes of the syntax tree, e. g. loops or instructions. Apart from information dealing with the test condition, loops consist of the loop body which is in turn represented by a syntax tree list. An *if*-statement is represented by three syntax tree lists representing the condition and the *then*- and *else*-parts. The one field of information attached to an instruction contains an expression tree.

In order to determine those parts of code lying between two occurrences $expr_{i,c}$ and $expr_{i,c+1}$ of a common subexpression $expr_i$, basically a bottom-up traversal of the syntax tree is performed. Starting from the tree nodes of type "instruction" containing $expr_{i,c}$ and $expr_{i,c+1}$, the tree is traversed towards the root of the tree representing the entire procedure. This process is stopped when

```
tree_node get_common_parent(tree_node n1, tree_node n2) {
  tree_node node, common_parent;

  /* Mark all nodes from expr_{i,c} (≈ n1) to root of tree */
  node = n1;
  while (node != ∅) {
    node.mark();
    node = node.parent(); }

  /* Starting from expr_{i,c+1} (≈ n2), find first marked node */
  node = n2;
  do
    node = node.parent();
  while (node.is_unmarked());
  common_parent = node;

  /* Remove all marks */
  node = n1;
  while (node != ∅) {
    node.unmark();
    node = node.parent(); }

  return common_parent; }
```

Figure 6.13. Algorithm to determine the Parent of $expr_{i,c}$ and $expr_{i,c+1}$

the paths starting from $expr_{i,c}$ and $expr_{i,c+1}$ pursued during the tree traversal join, i. e. when the first tree node that is both an ancestor of $expr_{i,c}$ and $expr_{i,c+1}$ is met. This traversal is represented by the dotted lines in figure 6.12 leading to the node labeled "Parent". Figure 6.13 depicts the algorithm used to determine this parent node. The trapezoid shaded area shown in figure 6.12 represents that part of the entire abstract syntax tree lying between $expr_{i,c}$ and $expr_{i,c+1}$.

The gathering of all instructions within this shaded area requires a second bottom-up traversal of the tree. Whenever a new parent node P is visited arriving from a child node C, the algorithm has to compute the position of C in the list of all sub-trees of P. Let pos_C denote this position of C. If P is reached on the path starting from $expr_{i,c}$ ($expr_{i,c+1}$ respectively), all sub-trees of P on the right of pos_C (on the left resp.) have to be considered. If P is the endpoint of the tree traversal where both paths join, only the sub-trees between these paths are relevant. The pseudo-code of this algorithm is given in figure 6.14.

In the beginning, the first common ancestor of $expr_{i,c}$ and $expr_{i,c+1}$ is determined and stored in common_parent. Hereafter, the path from $expr_{i,c}$ to node common_parent is traversed and all sub-trees on its right are added to the list

```
tree_list code_between(expr expr_{i,c}, expr expr_{i,c+1}) {
  tree_list result = ∅;
  tree_node node, common_parent, parent,
            n1 = get_node(expr_{i,c}), n2 = get_node(expr_{i,c+1});
  int lpos, rpos;

  /* Determine first common ancestor of expr_{i,c} and expr_{i,c+1} */
  common_parent = get_common_parent(n1, n2);

  /* Add all sub-trees right of first path to result */
  for (node=n1; node!=common_parent; node=parent) {
    parent = node.parent();
    lpos = parent.get_pos(node);
    if (parent != common_parent)
      for (all sub-trees T of parent with position > lpos)
        result.add(T); }

  /* Add all sub-trees left of second path to result */
  for (node=n2; node!=common_parent; node=parent) {
    parent = node.parent();
    rpos = parent.get_pos(node);
    if (parent != common_parent)
      for (all sub-trees T of parent with position < rpos)
        result.add(T); }

  /* Add sub-trees of common_parent between paths to result */
  for (all sub-trees T of common_parent with position > lpos and < rpos)
    result.add(T);

  return result; }
```

Figure 6.14.   Basic Algorithm for Syntax Tree Traversal

result. The same happens for the path starting from $expr_{i,c+1}$. Finally, the sub-trees of node common_parent lying between both paths need to be added.

This algorithm still lacks a correct handling of the relative positions of $expr_{i,c}$ and $expr_{i,c+1}$ to each other in the syntax tree. In particular, some possible constellations are not treated where the control flow is modified by loops and *if*-statements between two occurrences of an expression. This is stated more precisely in the following definition.

DEFINITION 6.5 (POSITIONS OF EXPRESSION OCCURRENCES)
Let $expr_{i,c}$ and $expr_{i,c+1}$ be two subsequent occurrences of an expression $expr_i$.

a *If **$expr_{i,c}$ occurs before a loop** containing $expr_{i,c+1}$, the entire body of the loop has to be taken as code lying between both occurrences of $expr_i$, as well as the code required for the loop's test condition.*

b *If **$expr_{i,c}$ occurs before an if-statement** containing $expr_{i,c+1}$, or $expr_{i,c}$ and $expr_{i,c+1}$ lie in the* then- *and* else-*part of the same if-statement, the entire if-statement including its* then- *and* else-*part has to be taken as code lying between both occurrences of $expr_i$, as well as its conditions.*

EXAMPLE 6.2

In the following, the different situations characterized in definition 6.5 are illustrated.

a *The left part of the following figure shows a code fragment where the expression* a * b *occurs immediately before a for-loop for the first time. The second occurrence of this expression is located in the middle of the loop's body. The structure of the corresponding syntax tree is depicted on the right hand side.*

```
...a*b...;
for (i=...)  {
    loop_body_1;
    ...a*b...;
    loop_body_2;
}
```

The algorithm shown in figure 6.14 would only put the code denoted as loop_body_1 *in the final list* result. *As a consequence, only this fragment of the loop body would be analyzed whether it modifies an operand of* a * b. *In the case that* a *or* b *are modified in* loop_body_2, *the analysis would lead to an incorrect result so that erroneous code is generated after the elimination of* a * b. *This can only be avoided by performing the analysis whether* a *or* b *are modified for the whole loop body. This argumentation is also valid for* do-while-*loops.*

In the situation that the first occurrence of a * b *is in a loop body and the second one after the loop, no particular care has to be taken. In this case, algorithm 6.14 puts all the code of the loop body after* a * b *in the list of code to be analyzed which is the correct behavior.*

b *The following code fragment depicts the situation where an expression occurs in the context of an* if-*statement and its corresponding representation by a syntax tree.*

```
1    ...a*b...;
2    if (...) {
3        then_part_1;
4        ...a*b...;
5        then_part_2;
6        ...a*b...;
7        then_part_3;
8    } else {
9        else_part_1;
10       ...a*b...;
11       else_part_2;
12       ...a*b...;
13       else_part_3;
14   }
```

For this code, it has to be decided whether the occurrence of a * b *in line 1 belongs to the same live range as the occurrences in the* then- *and* else-*parts of the* if-*statement. Since it can not be predicted which branch of the* if-*statement is executed, both branches have to be considered during the lifetime analysis.*

Due to this uncertainty, live ranges can not start outside an if-*statement and end within. Either, a live range has to start and stop inside one single branch of the* if-*statement, or it has to start before and to stop after the* if-*statement.*

Based on algorithm 6.14 and definition 6.5, the computation of maximum live ranges for expression occurrences can be presented. In the beginning, the very first occurrence of a given expression $expr_i$ is marked as the start of a live range.

Afterwards, all subsequent occurrences $expr_{i,c}$ and $expr_{i,c+1}$ are analyzed. Using algorithm 6.14, the code lying between these occurrences is determined. After this, the syntax tree is traversed one last time starting from $expr_{i,c+1}$. If a tree node representing a loop or an *if*-statement is visited during this traversal, a situation as illustrated in the figures of example 6.2 is detected. As a consequence, the entire loop or *if*-statement needs to be added to the code lying between $expr_{i,c}$ and $expr_{i,c+1}$.

All the parts of the syntax tree gathered during the previous phase are examined next. If code is detected modifying the operands of $expr_i$, the occurrences $expr_{i,c}$ and $expr_{i,c+1}$ do not belong to the same live range. As a consequence, $expr_{i,c}$ is marked as the end of a live range, and a new range starts at $expr_{i,c+1}$. Finally, the last occurrence $expr_{i,\#o_i}$ automatically is the end of a live range. This algorithm is shown as pseudo-code in figure 6.15.

In this algorithm, only a conservative estimation is performed whether $expr_i$ is modified by an expression E of sub_trees. Whenever it can not be precluded definitely that E modifies $expr_{i,c}$, E is assumed to modify $expr_i$. This especially holds whenever the address of an operand of $expr_i$ is taken and stored as a pointer. Since advanced code hoisting does not include a preprocessing

```
compute_live_ranges(expr expr_i) {
  tree_node node, loop = ∅, parent;

  start_of_range(expr_{i,1});

  /* Analyze all subsequent pairs of occurrences */
  for (c=2; c<=#o_i; c++) {
    tree_list sub_trees = code_between(expr_{i,c-1}, expr_{i,c});

    /* Check for situations of definition 6.5 */
    parent = get_common_parent(expr_{i,c-1}, expr_{i,c});
    for (node=get_node(expr_{i,c}); node!=parent; node=node.parent())
      if (node.is_loop())
        for (all sub-trees T of body & test condition of node)
          sub_trees.add(T);
      else if (node.is_if())
        for (all sub-trees T of test condition, then- & else-part of node)
          sub_trees.add(T);

    /* Mark current occurrences as start/end of range */
    for (all expressions E in sub_trees)
      if (E modifies expr_i) {
        end_of_range(expr_{i,c-1});
        start_of_range(expr_{i,c}); }
  }

  end_of_range(expr_{i,#o_i});
}
```

Figure 6.15. Algorithm to compute Live Ranges

step for alias analysis (see e. g. [Ryder et al., 2001]), every expression E manipulating a pointer of adequate type is assumed to modify $expr_i$ in such a situation. In the absence of pointers, it is sufficient to determine the variable v storing the result of an expression E. If v is used in $expr_i$, E modifies $expr_i$. This modification test is already provided by the SUIF API.

6.3.2 Determination of the outermost Loop for a CSE

As the final step, the positions in the code where a new variable *ach* has to be inserted have to be determined. This variable is used to store the value of an eliminated expression $expr_i$. This has to be done for all occurrences $expr_{i,j}$, ..., $expr_{i,k}$ lying in the same live range.

```
tree_node hoisted_position(expr_{i,j}, ..., expr_{i,k}) {
  tree_node P = get_node(expr_{i,j}), node;

  /* Determine first common parent of all occurrences */
  for (l=2; l<=k; l++)
    P = get_common_parent(P, get_node(expr_{i,l}));

  /* Move up tree, check for loops not modifying expr_i */
  node = P;
  while (node != ∅) {
    if (node.is_loop()) {
      bool is_loop_invariant = true;

      for (all expressions E of loop node)
        if (E modifies expr_i)
          is_loop_invariant = false;
      if (is_loop_invariant)
        P = node;
    }
    node = node.parent(); }
  return P; }
```

Figure 6.16.   Algorithm to determine the outermost Loop for a CSE

Since advanced code hoisting is not only based on common subexpressions but also on the optimized placement of loop-invariant code, this position is determined by looking for the outermost loop which is allowed to contain *ach*. In a first step, the first common ancestor P of all occurrences $expr_{i,j}, \ldots, expr_{i,k}$ is searched. P can be determined by calling the routine get_common_parent (see algorithm 6.13 on page 135) for every expression occurrence in a live range. This process is shown in the first part of algorithm 6.16.

From this position on, the syntax tree is traversed upwards to its root. Whenever a node is met representing a loop, it is checked whether this loop modifies the CSE $expr_i$. If this is not the case, $expr_i$ is invariant with respect to the current loop. As a consequence, the position P can be set to this loop. If a loop node modifies $expr_i$, P can not be set to the current node and thus remains unchanged as can be seen in the pseudo-code of algorithm 6.16. After the completion of this algorithm, P refers to the outermost position in the abstract syntax tree where a variable storing a common subexpression can be inserted.

6.3.3 Computation of Execution Frequencies using Polytope Models

As motivated in the introduction of this chapter, the most important property of advanced code hoisting is its criterion deciding when to eliminate an identified CSE. Based upon the exact calculation of execution frequencies of expressions, the elimination of a CSE is performed only if an expression is executed less frequently after the application of advanced code hoisting.

The example given in section 6.1 shows how control flow issues, namely loops and *if*-statements, influence the execution frequency of an expression. In order to compute execution frequencies, it is thus necessary to formally define the model of control flow supported by advanced code hoisting. For this purpose, definition 6.6 first specifies which kinds of loops are considered:

DEFINITION 6.6 (FOR-LOOPS AND AFFINE BOUNDS)
*A **for-loop L_l supported by advanced code hoisting** must fulfill the following requirements:*

a *The range of the index variable i_l of loop L_l always lies between the loop's lower and upper bounds lb_l and ub_l resp.: $lb_l \leq i_l \leq ub_l$ ($i_l \in \mathbb{Z}$).*

b *If loop L_l is nested in other for-loops $L_1, \ldots, L_{l-1}$, the lower and upper bounds can be affine expressions of the surrounding index variables $i_1, \ldots, i_{l-1}$. As a consequence, the index variable i_l iterates between*

$$lb_l = \sum_{j=1}^{l-1}(c_j' * i_j) + c' \ \leq \ i_l \ \leq \ \sum_{j=1}^{l-1}(c_j'' * i_j) + c'' = ub_l$$

for constants $c_j', c', c_j'', c'' \in \mathbb{Z}$.

c *After every iteration of L_l, i_l is incremented by a constant integer stride $s_l \in \mathbb{Z}$ with $s_l \neq 0$.*

The above definition of *for*-loops is mainly based on the fact that – in analogy to the techniques used for loop nest splitting presented in the previous chapter – the control flow of an embedded application is modeled using polytopes. Since polytopes are represented by linear equations and inequations (see definition 3.1

DEFINITION 6.7 (IF-STATEMENTS AND AFFINE CONDITIONS)
Let $L_1, \ldots, L_l$ denote a nest of for-loops complying with definition 6.6 which surrounds an if-statement.

a *An **if-statement supported by advanced code hoisting** has the format*

```
if  (C₁ ⊕ C₂ ⊕ ...)
    then_part;
else
    else_part;
```

where C_x are affine loop-variant conditions that are combined with logical operators $\oplus \in \{\&\&, | |\}$.

b *An **affine loop-variant condition** C_x is an affine expression of the index variables $i_1, \ldots, i_l$ of the surrounding loops. For constant values $c_j, c \in \mathbb{Z}$, C_x can be written as $C_x = \sum_{j=1}^{l}(c_j * i_j) \geq c$.*

The choice of the logical *AND* and *OR* operators in definition 6.7a is not a limitation since it was already shown in chapter 5 (see page 66) that this set of operators is fully sufficient to model all boolean combinations of affine conditions.

An expression *expr* which is identified to be an occurrence of a common subexpression can now be arbitrarily enclosed by control flow constructs according to definitions 6.6 and 6.7. An example source code structure is depicted in figure 6.17. As can be seen in this figure, *if*-statements are allowed to be nested in *for*-loops or in other *if*-statements. Additionally, *for*-loops may be guarded by nested *if*-statements. The number of nested *if*-statements within a given loop can vary between 0 and an arbitrary number denoted by m, n and o in the figure.

Whenever code is executed conditionally by virtue of an *if*-statement, it does not make a difference whether this code is located in the *if*-statement's *then*- or *else*-part since both cases can be treated by the techniques presented in the following. For this reason, all detailed information about *then*- and *else*-parts of *if*-statements is omitted in figure 6.17 for the sake of simplicity. This structure of control flow supported by advanced code hoisting is captured formally in the following definition.

DEFINITION 6.8 (GENERAL STRUCTURE OF CONTROL FLOW)
Let $\Lambda = \{L_1, \ldots, L_N\}$ be a nest of for-loops according to definition 6.6 and $\Upsilon = \{IF_1, \ldots, IF_O\}$ be a set of if-statements in accordance with definition 6.7. All loops and if-statements surround an expression $expr$.

*A sequence $\Gamma = (\gamma_1, \ldots, \gamma_M, \gamma_{M+1})$ of loops and if-statements is said to be a **nest of control flow structures** supported by advanced code hoisting if*

```
for (i₁ = c₁; i₁ ≤ c₂; i₁ += s₁)
    if (C₁,₁ ⊕ C₁,₂ ⊕ ...)

        ...

        if (Cₘ,₁ ⊕ Cₘ,₂ ⊕ ...)
            for (i₂ = lb₂; i₂ ≤ ub₂; i₂ += s₂)
                if (Cₘ₊₁,₁ ⊕ Cₘ₊₁,₂ ⊕ ...)

                    ...

                    if (Cₘ₊ₙ,₁ ⊕ Cₘ₊ₙ,₂ ⊕ ...)
                        for (iₙ = lbₙ; iₙ ≤ ubₙ; iₙ += sₙ)
                            if (Cₘ₊ₙ₊₁,₁ ⊕ Cₘ₊ₙ₊₁,₂ ⊕ ...)

                                ...

                                if (Cₘ₊ₙ₊ₒ,₁ ⊕ Cₘ₊ₙ₊ₒ,₂ ⊕ ...)
                                    expr;
```

Figure 6.17. General Structure of Control Flow surrounding an Expression *expr*

a *the first element of* Γ *is the outermost loop:* $\gamma_1 = L_1$,

b *the last element of* Γ *is the expression whose execution frequency is to be determined:* $\gamma_{M+1} = expr$,

c γ_m *either is a* for*-loop of* Λ *or an* if*-statement of* Υ *for* $2 \leq m \leq M$, *and*

d *every* if*-statement* $\gamma_m \in \Upsilon$ *only depends on the index variables* i_l *of surrounding loops* $L_l \in \{\gamma_1, \ldots, \gamma_{m-1}\} \cap \Lambda$.

For an if*-statement* $\gamma_m \in \Gamma$, TP_{γ_m} *denotes its* then*-part and* EP_{γ_m} *its* else*-part, respectively.*

In order to compute how many times an expression *expr* is executed under consideration of the code structure given in the above figure, a finite union of polytopes P^Γ is generated iteratively during a first phase. The goal of this phase is to build P^Γ such that it reflects the control flow given by the nested *for*-loops and *if*-statements. The constraints of P^Γ are generated according to the bounds of the loops and the conditions of the *if*-statements as will be shown in the following.

Since the only variables having influence on the control flow shown in figure 6.17 are the index variables of the loops $\Lambda = \{L_1, \ldots, L_N\}$ (since the *if*-statements only comprise affine conditions of the index variables' values), P^Γ is basically a subset of $\mathbb{Z}^N$: $P^\Gamma \subset \mathbb{Z}^N$. Every integer point included in P^Γ represents one single execution of expression *expr* for one actual assignment of values to the loops' index variables. The construction of P^Γ is specified in the following definition:

DEFINITION 6.9 (GENERAL POLYTOPE CONSTRUCTION)
Let $\Gamma = (\gamma_1, \ldots, \gamma_M, \gamma_{M+1})$ be a nest of control flow structures according to definition 6.8.
*The **finite union of polytopes** P^Γ modeling Γ is defined by*

$$P^\Gamma = (\bigcap_{\gamma_m \in \Lambda} P_m^{FOR}) \cap (\bigcap_{\gamma_m \in \Upsilon} P_m^{IF})$$

In order to construct a finite union of polytopes representing Γ, the finite unions of polytopes associated with the individual elements γ_m of Γ need to be intersected. Depending on whether γ_m is a *for*-loop or an *if*-statement, P_m^{FOR} or P_m^{IF} has to be taken in order to generate P^Γ. After its construction, P^Γ contains exactly all those values of the index variables $i_1, \ldots, i_N$ of Λ leading to the execution of the expression γ_{M+1}. This is proven later in this section.

For a given *for*-loop $\gamma_m \in \Lambda$, the following polyhedron P_m^{FOR} is generated:

DEFINITION 6.10 (POLYHEDRON FOR FOR-LOOPS)
Let $\Gamma = (\gamma_1, \ldots, \gamma_M, \gamma_{M+1})$ be a nest of control flow structures where Λ denotes the corresponding nested loops.
*For a loop $\gamma_m \in \Lambda$, a **polyhedron** P_m^{FOR} is created using the following constraints:*

a $i_m \geq \sum\limits_{j=1}^{m-1} (c_j' * i_j) + c'$ for lb_m

b $i_m \leq \sum\limits_{j=1}^{m-1} (c_j'' * i_j) + c''$ for ub_m

c $i_m = \begin{cases} \sum\limits_{j=1}^{m-1} (c_j' * i_j) + c' + (i_m' * s_m) & \text{for } s_m > 1, i_m' \in \mathbb{Z}, \\[2ex] \sum\limits_{j=1}^{m-1} (c_j'' * i_j) + c'' + (i_m' * s_m) & \text{for } s_m < -1, i_m' \in \mathbb{Z} \end{cases}$

Obviously, the constraints shown in definition 6.10a and 6.10b limit the iteration space of loop γ_m such that i_m can only range from $lb_m = \sum(c_j' * i_j) + c'$ to $ub_m = \sum(c_j'' * i_j) + c''$ for the constant values c_j', c', c_j'' and $c'' \in \mathbb{Z}$. These two constraints have the effect that i_m can take every integer value between lb_m and ub_m and thus model the case of a loop with a stride of 1 or -1.

For loops having a stride s_m other than 1 or -1, it is necessary to add a constraint in order to restrict i_m to only those integer values between lb_m and ub_m that can be reached using s_m. In the case of a stride $s_m > 1$, i.e. a loop counting forwards from lb_m up to ub_m, the constraint $i_m = lb_m + (i_m' * s_m)$ is added to P_m^{FOR}. Here, i_m' is an auxiliary integer variable for loop γ_m. This constraint has the effect that the index variable i_m can take any value being a multiple of the stride s_m which is added to the lower loop bound lb_m. Starting with lb_m itself, only every s_m-th integer value can be assigned to i_m. In the case of a negative loop stride, i.e. a loop counting backwards from ub_m down to lb_m, the constraint $i_m = ub_m + (i'_m * s_m)$ is used. In this case, the multiples of the negative value s_m are added to ub_m so that only every $|s_m|$-th value smaller than ub_m is assigned to i_m.

It is not necessary to add constraints for i'_m since this is already done implicitly by the constraints modeling the lower and upper loop bounds. In the case of a positive stride, $i_m \geq lb_m$ implies $i'_m \geq 0$, and $i_m \leq ub_m$ necessarily defines an upper bound for i'_m. The same holds for $s_m < 0$: here, $i_m \leq ub_m$ implies $i'_m \geq 0$, and an upper bound for i'_m is implicitly defined by $i_m \geq lb_m$.

EXAMPLE 6.3

The application of definition 6.10 to the following for-*loop*

```
for (i=3; i<19; i+=4)
```

leads to the polytope

$$P_i^{FOR} \;=\; \{\,(i \geq 3) \wedge (i \leq 18) \wedge (i = 3 + (i' * 4))\,\}$$

i is only allowed to take those values which are a multiple of 4 added to the lower bound of 3, and which lie between 3 and 18. By means of this example, it can be seen that constraints for i' are not necessary. i' is unable to hold a negative value since the third constraint would lead to the situation that i is assigned a value resulting from the addition of a negative number to the lower bound. This violates the first constraint ensuring that i is greater than or equal to the lower loop bound. On the other hand, i' can not be greater than 3 since this would result in the assignment of a value greater than or equal to 19 to i violating the second constraint $i \leq 18$.

In their combination, the constraints of P_i^{FOR} ensure that i can only take the values 3, 7, 11 and 15 which represents exactly the iterations of the above for-*loop. Assuming that the stride of the above loop is changed to a negative value*

```
for (j=18; j>=3; j-=4)
```

the associated polytope P_j^{FOR} differs from P_i^{FOR} only with regard to the third constraint:

$$P_j^{FOR} \;=\; \{\,(j \geq 3) \wedge (j \leq 18) \wedge (j = 18 + (j' * -4))\,\}$$

In this case also, P_j^{FOR} only consists of those values assigned to j by the loop, namely 18, 14, 10 and 6.

THEOREM 6.1 (CORRECTNESS OF P_m^{FOR})

For a given nest Γ of control flow structures, let $\gamma_m \in \Lambda$ be a for-loop in compliance with definition 6.6. Let P_m^{FOR} be the corresponding polyhedron according to definition 6.10.

P_m^{FOR} *contains exactly those values of the index variable i_m for which γ_{m+1} is executed.*

Proof

Let $V = (v_1, \ldots, v_z)$ be the sequence of values assigned to the index variable i_m during the execution of loop γ_m. To be shown per complete induction on $x \in \{1, \ldots, z\}$:
$$\forall v_x : v_x \in V \Leftrightarrow v_x \in P_m^{FOR}$$

Case 1: Positive loop stride $s_m > 0$

x = 1:

$v_1 \in V$

$\Leftrightarrow (v_1 = lb_m) \wedge (v_1 \leq ub_m)$ *(due to initialization of i_m*
                                                    with lb_m for first iteration)

$\Leftrightarrow \left(v_1 = \sum_{j=1}^{m-1} (c_j' * i_j) + c' \right) \wedge \left(v_1 \leq \sum_{j=1}^{m-1} (c_j'' * i_j) + c'' \right)$

$\Leftrightarrow \left(v_1 = \sum_{j=1}^{m-1} (c_j' * i_j) + c' + (0 * s_m) \right) \wedge \left(v_1 \leq \sum_{j=1}^{m-1} (c_j'' * i_j) + c'' \right)$

$\Leftrightarrow v_1$ fulfills definitions 6.10a, 6.10b and 6.10c

$\Leftrightarrow v_1 \in P_m^{FOR}$

x → x+1:

$v_{x+1} \in V$

$\Leftrightarrow (v_{x+1} \geq lb_m) \wedge (v_{x+1} \leq ub_m) \wedge (v_{x+1} = v_x + s_m)$

$\Leftrightarrow \left(v_{x+1} \geq \sum_{j=1}^{m-1} (c_j' * i_j) + c' \right) \wedge \left(v_{x+1} \leq \sum_{j=1}^{m-1} (c_j'' * i_j) + c'' \right) \wedge$

$\left(v_{x+1} = v_x + s_m \right)$

The induction's presumption $v_x \in V \Leftrightarrow v_x \in P_m^{FOR}$ yields

$$v_x \in P_m^{FOR} \Rightarrow v_x = \sum_{j=1}^{m-1} (c_j' * i_j) + c' + (i_m' * s_m)$$

due to definition 6.10c. The application of this presumption to v_{x+1} leads to:

$\Leftrightarrow \left(v_{x+1} \geq \sum_{j=1}^{m-1} (c_j' * i_j) + c' \right) \wedge \left(v_{x+1} \leq \sum_{j=1}^{m-1} (c_j'' * i_j) + c'' \right) \wedge$

$\left(v_{x+1} = \sum_{j=1}^{m-1} (c_j' * i_j) + c' + (i_m' * s_m) + s_m \right)$

$\Leftrightarrow \left(v_{x+1} \geq \sum_{j=1}^{m-1} (c_j' * i_j) + c' \right) \wedge \left(v_{x+1} \leq \sum_{j=1}^{m-1} (c_j'' * i_j) + c'' \right) \wedge$

$\left(v_{x+1} = \sum_{j=1}^{m-1} (c_j' * i_j) + c' + (i_m' + 1) * s_m \right)$

$\Leftrightarrow v_x$ fulfills definitions 6.10a, 6.10b and 6.10c

$$\Leftrightarrow v_x \in P_m^{FOR}$$

Case 2: Negative loop stride $s_m < 0$

Analogously to the scheme shown above.

$\square$

For a given *if*-statement $\gamma_m \in \Upsilon$, the following finite union of polyhedra P_m^{IF} is generated:

DEFINITION 6.11 (UNION OF POLYHEDRA FOR IF-STATEMENTS)
Let $\Gamma = (\gamma_1, \ldots, \gamma_M, \gamma_{M+1})$ *be a nest of control flow structures composed of a loop nest* Λ *and a set of* if-statements Υ.

For an if-statement $\gamma_m = (C_1 \oplus C_2 \oplus \ldots \oplus C_n) \in \Upsilon$ *surrounded by loops* $L_1, \ldots, L_l$, π *denotes the permutation of* $\{1, \ldots, n\}$ *representing the natural execution order of the conditions* C_x.

a *For an affine condition* $C_x = \sum_{j=1}^{l} (c_j * i_j) \geq c$ *of* γ_m, *the following polyhedra* P_x *and* $\overline{P_x}$ *are required:*

$$
\begin{aligned}
P_x &= \{ (i_1, \ldots, i_l) \in \mathbb{Z}^l \mid c_1 * i_1 + \ldots + c_l * i_l \geq c \} \\
\overline{P_x} &= \{ (i_1, \ldots, i_l) \in \mathbb{Z}^l \mid -c_1 * i_1 - \ldots - c_l * i_l \geq -c + 1 \}
\end{aligned}
$$

b *In order to represent the first condition* $C_{\pi(1)}$ *of* γ_m, *the polyhedron* P_x *is used if the next element* γ_{m+1} *of* Γ *is located in the* then*-part of* γ_m *(*$\overline{P_x}$ *for the* else*-part, respectively).*

$$
P_{m,1}^{IF} = \begin{cases} P_{\pi(1)} & \text{for } \gamma_{m+1} \in TP_{\gamma_m} \\ \overline{P_{\pi(1)}} & \text{for } \gamma_{m+1} \in EP_{\gamma_m} \end{cases}
$$

c *For the remaining conditions* $C_{\pi(x)}$ *with* $x \in \{2, \ldots, n\}$, *the corresponding finite union of polytopes* $P_{m,x}^{IF}$ *on the one hand depends on the logical operator used to combine* $C_{\pi(x)}$ *with* $C_{\pi(x-1)}$. *On the other hand, the position of* γ_{m+1} *in the* then- *or* else*-part of* γ_m *has to be considered, too:*

$$
P_{m,x}^{IF} = \begin{cases} P_{m,x-1}^{IF} \cap P_{\pi(x)} & \text{for } \gamma_{m+1} \in TP_{\gamma_m} \text{ and } C_{\pi(x-1)} \text{ \&\& } C_{\pi(x)}, \\ P_{m,x-1}^{IF} \cup P_{\pi(x)} & \text{for } \gamma_{m+1} \in TP_{\gamma_m} \text{ and } C_{\pi(x-1)} \text{ || } C_{\pi(x)}, \\ P_{m,x-1}^{IF} \cup \overline{P_{\pi(x)}} & \text{for } \gamma_{m+1} \in EP_{\gamma_m} \text{ and } C_{\pi(x-1)} \text{ \&\& } C_{\pi(x)}, \\ P_{m,x-1}^{IF} \cap \overline{P_{\pi(x)}} & \text{for } \gamma_{m+1} \in EP_{\gamma_m} \text{ and } C_{\pi(x-1)} \text{ || } C_{\pi(x)} \end{cases}
$$

d *The* **finite union of polyhedra** P_m^{IF} *representing the entire* if-statement γ_m *is defined as*

$$P_m^{IF} = P_{m,n}^{IF}$$

As can be seen from definition 6.11, P_m^{IF} is generated iteratively. In order to build P_m^{IF}, the polyhedral representation of every condition C_x is required (see definition 6.11a). For every condition of γ_m, the polyhedra P_x directly representing C_x and $\overline{P_x}$ representing the negated condition $\overline{C_x}$ are needed.

Definition 6.11b shows that $P_{m,1}^{IF}$ is equal to the polyhedron representing the first condition of γ_m according to the execution order π. During this step, the position of the next element $\gamma_{m+1} \in \Gamma$ in relation to the *if*-statement γ_m has to be considered. If γ_{m+1} is located in the *then*-part of γ_m, it is executed only if γ_m is satisfied so that $P_{\pi(1)}$ has to be taken. In contrast, the placement of γ_{m+1} in the *else*-part of γ_m implies the choice of $\overline{P_{\pi(1)}}$ since γ_{m+1} is executed only if γ_m is not satisfied.

After this, all remaining conditions C_x of γ_m are examined using the ordering given by π. For every condition C_x, the finite union of polyhedra $P_{m,x-1}^{IF}$ generated so far is connected with the appropriate polyhedron representing C_x. In analogy to the previous paragraph, the polyhedron $P_{\pi(x)}$ is used if γ_{m+1} is located in the *then*-part of γ_m (the first two cases of definition 6.11c), or $\overline{P_{\pi(x)}}$ for $\gamma_{m+1} \in EP_{\gamma_m}$ (last two cases of definition 6.11c), respectively.

If $P_{\pi(x)}$ is taken, the intersection of polyhedra is applied whenever conditions $C_{\pi(x)}$ and $C_{\pi(x-1)}$ are connected using the logical *AND* operator (first case of definition 6.11c). For the logical *OR*, the union of polyhedra is used (second case). For the loop or *if*-statement γ_{m+1} being located in the *else*-part of γ_m, the negation of the conditions combined with the application of de Morgan's rules has the effect that the union has to be used for the logical *AND* (third case) and vice versa (fourth case).

EXAMPLE 6.4

Let C_1 denote the condition 4*i + j >= 12 *and C_2 represent the condition* i + 5*j <= 28 *for some index variables* i *and* j. *The application of definition 6.11a to C_1 leads to the polyhedra*

$$
\begin{aligned}
P_1 &= \{\,(i,j) \in \mathbb{Z}^2 \mid 4*i + j \geq 12\,\} \\
\overline{P_1} &= \{\,(i,j) \in \mathbb{Z}^2 \mid -4*i - j \geq -11\,\}
\end{aligned}
$$

where P_1 obviously represents C_1 and $\overline{P_1}$ denotes the negated condition $\overline{C_1}$. Analogously, the following polyhedra are defined for condition C_2:

$$
P_2 = \{\,(i,j) \in \mathbb{Z}^2 \mid -i - 5*j \geq -28\,\}
$$

which is the straightforward case since the polyhedra representing the conditions simply need to be intersected in order to model the logical AND. *Similarly, the union operator has to be used for*

    if $(C_1 \; || \; C_2)$ *expr*;

since in this case, all values of i *and* j *have to be captured by* P^{IF} *which belong either to* P_1 *or* P_2:

$$P^{IF} \;\; = \;\; P_1 \cup P_2 \;\; =$$
$$\{ (i,j) \in \mathbb{Z}^2 \mid 4 * i + j \geq 12 \} \cup \{ (i,j) \in \mathbb{Z}^2 \mid -i - 5 * j \geq -28 \}$$

Obviously, the if-statement if $(C_1 \; \&\& \; C_2)$ ... else *expr; is equivalent to* if $!(C_1 \; \&\& \; C_2)$ *expr;* else ... *De Morgan's rule applied to the conditions of the if-statement thus leads to* if $(!C_1 \; || \; !C_2)$ *expr;* else ...

As a consequence, it is obvious that P^{IF} *consists of the negated polyhedra* $\overline{P_1}$ *and* $\overline{P_2}$. *Furthermore, the use of the union operator for polyhedra in this situation can be explained with the switching of the logical operator from* AND *to* OR *due to de Morgan's rule. All in all, the following finite union of polyhedra corresponds to the above if-statement:*

$$P^{IF} \;\; = \;\; \overline{P_1} \cup \overline{P_2} \;\; =$$
$$\{ (i,j) \in \mathbb{Z}^2 \mid -4 * i - j \geq -11 \} \cup \{ (i,j) \in \mathbb{Z}^2 \mid i + 5 * j \geq 29 \}$$

Analogously, the if-statement

    if $(C_1 \; || \; C_2)$ ... else *expr*;

is represented correctly by

$$P^{IF} \;\; = \;\; \overline{P_1} \cap \overline{P_2} \;\; = \;\; \{ (i,j) \in \mathbb{Z}^2 \mid (-4 * i - j \geq -11) \wedge (i + 5 * j \geq 29) \}$$

THEOREM 6.2 (CORRECTNESS OF P_m^{IF})

For a given nest Γ *of control flow structures, let* $\gamma_m = (C_1 \oplus \ldots \oplus C_n) \in \Upsilon$ *be an if-statement in compliance with definition 6.7. Let* π *denote the execution order of* γ_m *and* P_m^{IF} *be the corresponding finite union of polyhedra according to definition 6.11.*

P_m^{IF} contains exactly those values of the index variables of Λ *for which* γ_{m+1} *is executed.*

Proof

The proof is structured in two parts which separately deal with the cases that γ_{m+1} is located in the *then-* or *else*-part of the *if*-statement γ_m.

Let $L_1, \ldots, L_l$ be the loops surrounding γ_m and $i = (i_1, \ldots, i_l)$ be an assignment of values to their index variables.

Case 1: $\gamma_{m+1} \in TP_{\gamma_m}$
$\Leftrightarrow \gamma_{m+1}$ is executed only if $(C_1 \oplus \ldots \oplus C_n) = true$.

To be shown per complete induction on $x \in \{1, \ldots, n\}$:
$C_1(i) \oplus \ldots \oplus C_n(i) = true \Leftrightarrow i \in P_m^{IF} = P_{m,n}^{IF}$

x = 1:

$$C_{\pi(1)}(i) = true$$

$$\Leftrightarrow \sum_{j=1}^{l} c_j * i_j \geq c \qquad\qquad\qquad \text{(due to definition 6.7b)}$$

$$\Leftrightarrow c_1 * i_1 + \ldots + c_l * i_l \geq c$$

$$\Leftrightarrow (i_1, \ldots, i_l) \in P_{\pi(1)} = P_m^{IF} \qquad \text{(due to definition 6.11a for } C_{\pi(1)})$$

x → x+1:

For the logical *AND*:

$$(C_{\pi(1)}(i) \oplus \ldots \oplus C_{\pi(x)}(i)) \,\&\&\, C_{\pi(x+1)}(i) = true$$

$$\Leftrightarrow C_{\pi(1)}(i) \oplus \ldots \oplus C_{\pi(x)}(i) = true \text{ and } C_{\pi(x+1)}(i) = true$$

To the left part of the above line, the presumption of the induction is applied. W. r. t. the right part, it is obvious to see that $C_{\pi(x+1)}(i) = true$ is equivalent to $(i_1, \ldots, i_l) \in P_{\pi(x+1)}$ due to definition 6.11a.

$$\Leftrightarrow (i_1, \ldots, i_l) \in P_{m,x}^{IF} \text{ and } (i_1, \ldots, i_l) \in P_{\pi(x+1)}$$

$$\Leftrightarrow (i_1, \ldots, i_l) \in P_{m,x}^{IF} \cap P_{\pi(x+1)}$$

Since the last line denotes P_m^{IF} for the $x + 1$st step (see case 1 of definition 6.11c), $i \in P_m^{IF}$ holds.

For the logical *OR*:

$$(C_{\pi(1)}(i) \oplus \ldots \oplus C_{\pi(x)}(i)) \,||\, C_{\pi(x+1)}(i) = true$$

$$\Leftrightarrow C_{\pi(1)}(i) \oplus \ldots \oplus C_{\pi(x)}(i) = true \text{ or } C_{\pi(x+1)}(i) = true$$

Here again, the presumption of the induction and definition 6.11a are applied.

$$\Leftrightarrow (i_1, \ldots, i_l) \in P_{m,x}^{IF} \text{ or } (i_1, \ldots, i_N) \in P_{\pi(x+1)}$$

$$\Leftrightarrow (i_1, \ldots, i_l) \in P_{m,x}^{IF} \cup P_{\pi(x+1)}$$

Since the last line denotes P_m^{IF} for the $x + 1$st step (see case 2 of definition 6.11c), $i \in P_m^{IF}$ holds.

Case 2: $\gamma_{m+1} \in EP_{\gamma_m}$

$\Leftrightarrow \gamma_{m+1}$ is executed only if $!(C_1(i) \oplus \ldots \oplus C_n(i)) = true$.

To be shown per complete induction on $x \in \{1, \ldots, n\}$:
$$!(C_1(i) \oplus \ldots \oplus C_n(i)) = true \Leftrightarrow i \in P_m^{IF} = P_{m,n}^{IF}$$

x = 1:

$$!(C_{\pi(1)}(i)) = true$$

$$\Leftrightarrow !(\sum_{j=1}^{l} c_j * i_j \geq c) \qquad\qquad\qquad \text{(due to definition 6.7b)}$$

$$\Leftrightarrow \sum_{l=1}^{l} c_j * i_j < c$$

$$\Leftrightarrow c_1 * i_1 + \ldots + c_j * i_j < c$$

$$\Leftrightarrow -c_1 * i_1 - \ldots - c_l * i_l \geq -c + 1$$

$$\Leftrightarrow (i_1, \ldots, i_l) \in \overline{P_{\pi(1)}} = P_m^{IF} \qquad (\textit{due to definition 6.11a for } C_{\pi(1)})$$

x → x+1:

For the logical *AND*:

$$!((C_{\pi(1)}(i) \oplus \ldots \oplus C_{\pi(x)}(i)) \text{ \&\& } C_{\pi(x+1)}(i)) = \textit{true}$$

$$\Leftrightarrow (!(C_{\pi(1)}(i) \oplus \ldots \oplus C_{\pi(x)}(i)) \text{ || } !C_{\pi(x+1)}(i)) = \textit{true}$$

$$\Leftrightarrow !(C_{\pi(1)}(i) \oplus \ldots \oplus C_{\pi(x)}(i)) = \textit{true} \text{ or } !C_{\pi(x+1)}(i) = \textit{true}$$

The application of the induction's presumption and definition 6.11a leads to:

$$\Leftrightarrow (i_1, \ldots, i_l) \in P_{m,x}^{IF} \text{ or } (i_1, \ldots, i_l) \in \overline{P_{\pi(x+1)}}$$

$$\Leftrightarrow (i_1, \ldots, i_l) \in P_{m,x}^{IF} \cup \overline{P_{\pi(x+1)}}$$

Since the last line denotes P_m^{IF} for the $x + 1$st step (see case 3 of definition 6.11c), $i \in P_m^{IF}$ holds.

For the logical *OR*:

$$!((C_{\pi(1)}(i) \oplus \ldots \oplus C_{\pi(x)}(i)) \text{ || } C_{\pi(x+1)}(i)) = \textit{true}$$

$$\Leftrightarrow (!(C_{\pi(1)}(i) \oplus \ldots \oplus C_{\pi(x)}(i)) \text{ \&\& } !C_{\pi(x+1)}(i)) = \textit{true}$$

$$\Leftrightarrow !(C_{\pi(1)}(i) \oplus \ldots \oplus C_{\pi(x)}(i)) = \textit{true} \text{ and } !C_{\pi(x+1)}(i) = \textit{true}$$

The application of the induction's presumption and definition 6.11a leads to:

$$\Leftrightarrow (i_1, \ldots, i_l) \in P_{m,x}^{IF} \text{ and } (i_1, \ldots, i_l) \in \overline{P_{\pi(x+1)}}$$

$$\Leftrightarrow (i_1, \ldots, i_l) \in P_{m,x}^{IF} \cap \overline{P_{\pi(x+1)}}$$

Since the last line denotes P_m^{IF} for the $x + 1$st step (see case 4 of definition 6.11c), $i \in P_m^{IF}$ holds.

$\square$

Now that the correctness of the polytopes representing *for*-loops and *if*-statements is proven, it is only left to be shown that the finite union of polytopes P^Γ (see definition 6.9 on page 144) reflects Γ accurately.

THEOREM 6.3 (CORRECTNESS OF P^Γ)

Let $\Gamma = (\gamma_1, \ldots, \gamma_M, \gamma_{M+1})$ *be a sequence of control flow structures composed of a loop nest* Λ *and a set of if-statements* Υ *(see definition 6.8). Furthermore, let* P^Γ *be the finite union of polytopes associated to* Γ *according to definition 6.9.*

P^Γ contains exactly those values of the index variables of Λ *for which the expression expr* $= \gamma_{M+1}$ *is executed.*

Proof

Expression γ_{M+1} is executed exactly for all values of the index variables of $\Lambda = \{L_1, \ldots, L_N\}$ which

- lie within all loop bounds and are reached for the given strides, and
- satisfy all *if*-statements of Υ^2.

For every execution of γ_{M+1}, the index variables $i = (i_1, \ldots, i_N)$ must fulfill the following formal conditions:

$$\forall i_l \ (1 \leq l \leq N) : lb_l \leq i_l \leq ub_l \ \wedge$$

$$\forall i_l \ (1 \leq l \leq N) : i_l = b_l + (i'_l * s_l) \text{ with } b_l = \begin{cases} lb_l & \text{for } s_l > 0 \\ ub_l & \text{for } s_l < 0 \end{cases} \ \wedge$$

$$\forall \gamma_m \in \Upsilon : \gamma_m \text{ is satisfied for } i_1, \ldots, i_N$$

For the sake of simplicity, the following notation for the three conditions listed above is used in the remainder of this proof:

$$(lb \leq i \leq ub) \ \wedge \ (i = b + (i' * s)) \ \wedge \ (\Upsilon(i) \text{ is satisfied})$$

Using complete induction on the depth M of the entire nest Γ of control flow structures, it is shown that:

$$(lb \leq i \leq ub) \ \wedge \ (i = b + (i' * s)) \ \wedge \ (\Upsilon(i) \text{ is satisfied}) \Leftrightarrow i \in P^\Gamma$$

M = 1:

Due to definition 6.8a, γ_1 must be the *for*-loop L_1. As a consequence, the set of *if*-statements Υ is empty. Hence, for $i = (i_1)$ it holds:

$$(lb \leq i_1 \leq ub) \ \wedge \ (i_1 = b + (i'_1 * s_1))$$
$$\Leftrightarrow i_1 \in P_1^{FOR} \hspace{3cm} \textit{(due to theorem 6.1)}$$
$$\Leftrightarrow i \in P^\Gamma \hspace{2cm} \textit{(since } P^\Gamma = P_1^{FOR}\textit{, definition 6.9)}$$

M → M+1:

Case 1: $\gamma_{M+1} \in \Lambda$
Let $i^M = (i_1, \ldots, i_M)$ denote the values of the index variables without the new item i_{M+1} (lb^M, ub^M, b^M, i'^M and s^M analogously).

$$(lb \leq i \leq ub) \ \wedge \ (i = b + (i' * s)) \ \wedge \ (\Upsilon(i) \text{ is satisfied})$$
$$\Leftrightarrow (lb^M \leq i^M \leq ub^M)' \wedge \ (i^M = b^M + (i'^M * s^M)) \ \wedge$$
$$(lb_{M+1} \leq i_{M+1} \leq ub_{M+1}) \ \wedge \ (i_{M+1} = b_{M+1} + (i'_{M+1} * s_{M+1}))$$
$$\wedge \ (\Upsilon(i) \text{ is satisfied})$$

---

[2]In this proof, it is not considered whether γ_{m+1} is located in the *then*- or *else*-part of an *if*-statement γ_m. As a consequence, it is not differentiated here whether the negated conditions of γ_m or the non-negated conditions need to be satisfied. In order to keep the proof simple, only the abstract satisfaction of an *if*-statement γ_m is dealt with here.

$$\Leftrightarrow (lb^M \le i^M \le ub^M) \wedge (i^M = b^M + (i'^M * s^M)) \wedge$$
$$(lb_{M+1} \le i_{M+1} \le ub_{M+1}) \wedge (i_{M+1} = b_{M+1} + (i'_{M+1} * s_{M+1}))$$
$$\wedge (\Upsilon(i^M) \text{ is satisfied})$$

The last equivalence holds since Υ remains unchanged during the consideration of a new loop γ_{M+1} in this case. As a consequence, Υ can not contain any *if*-statement depending on the new index variable i_{M+1} so that Υ necessarily must be satisfied for i^M.

The application of the induction's presumption leads to:

$$\Leftrightarrow (i^M \in P^\Gamma) \wedge$$
$$(lb_{M+1} \le i_{M+1} \le ub_{M+1}) \wedge (i_{M+1} = b_{M+1} + (i'_{M+1} * s_{M+1}))$$
$$\Leftrightarrow (i^M \in P^\Gamma) \wedge (i_{M+1} \in P^{FOR}_{M+1}) \qquad \textit{(due to theorem 6.1)}$$
$$\Leftrightarrow i = (i_1, \ldots, i_M, i_{M+1}) \in P^\Gamma \cap P^{FOR}_{M+1}$$

This final equivalence is valid since P^Γ and P^{FOR}_{M+1} are defined on disjoint domains of values. P^Γ is only defined on the index variables $i_1, \ldots, i_M$ and does not constrain i_{M+1}. In contrast, P^{FOR}_{M+1} only restricts the domain of values of i_{M+1}, but not of $i_1, \ldots, i_M$. As a consequence, $P^\Gamma \cap P^{FOR}_{M+1}$ contains all the points of P^Γ – extended by the new domain i_{M+1} – that simultaneously satisfy the constraints of P^{FOR}_{M+1}.

Case 2: $\gamma_{M+1} \in \Upsilon$
Let Υ' denote the set of *if*-statements without the new item γ_{M+1}.

$$(lb \le i \le ub) \wedge (i = b + (i' * s)) \wedge (\Upsilon(i) \text{ is satisfied})$$
$$\Leftrightarrow (lb \le i \le ub) \wedge (i = b + (i' * s)) \wedge$$
$$(\Upsilon'(i) \text{ is satisfied}) \wedge (\gamma_{M+1}(i) \text{ is satisfied})$$
$$\Leftrightarrow (i \in P^\Gamma) \wedge (\gamma_{M+1}(i) \text{ is satisfied}) \qquad \textit{(induction's presumption)}$$
$$\Leftrightarrow (i \in P^\Gamma) \wedge (i \in P^{IF}_{M+1}) \qquad \textit{(due to theorem 6.2)}$$
$$\Leftrightarrow i \in P^\Gamma \cap P^{IF}_{M+1}$$

$\square$

Theorem 6.3 is helpful for the computation of the execution frequency of an expression *expr* since it states that P^Γ contains exactly one point $i \in \mathbb{Z}^N$ for every time *expr* is executed. As a consequence, the execution frequency of *expr* is equal to the size of P^Γ.

COROLLARY 6.1
Let $\Gamma = (\gamma_1, \ldots, \gamma_M, \gamma_{M+1})$ be a sequence of control flow structures complying with definition 6.8 and P^Γ be the corresponding finite union of polytopes according to definition 6.9.

The execution frequency of the expression $expr = \gamma_{M+1}$ *is equal to the size of* P^Γ:

$$\#\,expr \;=\; |P^\Gamma|$$

As already stated in chapter 3.1, the computation of the number of points included in a finite union of polytopes is #*P*-complete [Kaibel and Pfetsch, 2003]. In order to determine the execution frequency of an expression *expr*, the techniques described in [Clauss and Loechner, 1998] for the computation of a polytope's size are applied to P^Γ. A detailed description of these techniques is omitted here since it is beyond the scope of this book. In short, the parametric vertices are computed based on the linear constraints of P^Γ in a first step. Using these vertices, the so called *Ehrhart polynomial* is determined which is a parametric representation of the number of integer points of P^Γ. For more details, the interested reader is referred to [Clauss and Loechner, 1998].

Due to the #*P*-completeness of the techniques mentioned above, the computation of the execution frequency of an expression *expr* does not have a polynomial complexity. Instead, the worst-case complexity of the techniques described in this section is exponential. However, as will be shown in the following section 6.4, the use of this methodology for real-life applications leads to feasibly short runtimes of only a few CPU seconds.

6.4   Experimental Results

The techniques described in the previous section are fully automated using the SUIF intermediate format [Wilson et al., 1995] and the polyhedral library Polylib [Loechner, 1999].

In this section, detailed experimental results for advanced code hoisting are provided. For this purpose, the optimization is applied to the source codes of the CAVITY [Bister et al., 1989] and QSDPCM [Strobach, 1988] algorithms which have already been described in section 5.5.1 (see page 101). The efficiency of the polyhedral analysis employed for advanced code hoisting is apparent by virtue of the low runtimes required for the optimization of these benchmarks. Using a Pentium IV based host machine running at 2.6 GHz, the entire optimization and transformation of CAVITY requires only 0.44 CPU seconds. In the case of QSDPCM, 29.9 seconds are necessary.

The results given in the following include the measurement of pipeline and cache behavior (see section 6.4.1), execution times and code sizes (section 6.4.2) and energy dissipation (section 6.4.3). The numerical values corresponding to the subsequent diagrams can be found in appendix C.

6.4.1   Pipeline and Cache Performance

In analogy to the previous chapter 5 dealing with loop nest splitting, the presentation of the experimental results for advanced code hoisting starts with

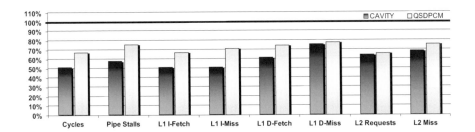

Figure 6.18. Relative Pipeline and Cache Behavior for Intel Pentium III

the detailed description of the instruction pipeline and cache behavior. Again, the hardware counters of an Intel Pentium III processor, a Sun UltraSPARC III and a MIPS R10000 were monitored for this purpose.

Intel Pentium III

The influence of advanced code hoisting on the behavior of the instruction pipeline and on first / second level instruction and data caches of the Intel Pentium III is depicted in figure 6.18. The 100% base line highlighted in bold denotes the benchmarks' behavior before the optimization. The bars of the diagram represent their behavior after advanced code hoisting relative to this base line.

As can be seen from this figure, advanced code hoisting is very beneficial since it leads to speed-ups of 32.7% for QSDPCM up to 49.1% for CAVITY (see column Cycles). These speed-ups are mainly due to the improved reuse of computed results and the inherent reduction of instruction executions. This is clearly shown by the fact that the measured speed-ups are of exactly the same order of magnitude as the number of instruction fetches (column L1 I-Fetch). As a consequence, advanced code hoisting leads to less L1 I-cache misses which are reduced between 28.5% (QSDPCM) and 49.1% (CAVITY) by the optimization.

Since already computed results are reused after advanced code hoisting, it is not surprising that less data fetches are measured since the operands of the eliminated expressions are accessed less frequently. This is illustrated in column L1 D-Fetch where reductions of data accesses between 25.6% and 38.5% are shown. These reductions come along with a drop of data cache misses between 22.2% and 24.4% (column L1 D-Miss).

The improved behavior of the benchmarks after optimization w. r. t. the first level caches also causes a benefit in the second level unified cache. As can be seen from columns L2 Requests and L2 Miss, the number of cache accesses goes

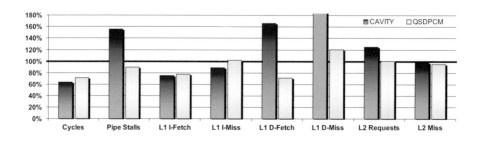

Figure 6.19.   Relative Pipeline and Cache Behavior for Sun UltraSPARC III

down by 34.1% and 35.8% leading to reductions of L2 cache misses between 24% and 31.4%.

The results for the Intel Pentium III show that advanced code hoisting leads to a higher locality of both instructions and data for this platform. Due to the reduced numbers of instruction and data fetches, the processor executes less wait states as shown in column Pipe Stall. Reductions ranging from 24.7% (QSDPCM) up to 42.5% (CAVITY) were achieved by advanced code hoisting.

Sun UltraSPARC III

The results for the Sun UltraSPARC III processor also show a significant speed-up after advanced code hoisting. Column Cycles of figure 6.19 illustrates that gains between 29% (QSDPCM) and 36.4% (CAVITY) were achieved in this case. Similar to the situation of the Intel Pentium processor, these improvements stem from the reduced number of instruction executions after this source code optimization. This is shown by means of columns L1 I-Fetch and L1 I-Miss, where the amount of instruction fetches is reduced by 22.7% (QSDPCM) – 24.5% (CAVITY). In the case of QSDPCM, the number of instruction cache misses remains constant after the optimization, but an improvement of 11% was measured for the CAVITY benchmark.

Column L1 D-Fetch shows for the CAVITY benchmark that advanced code hoisting increases the number of data fetches by 66.6%. This behavior is due to the fact that the Sun compiler is unable to use the processor's registers as efficiently after the optimization as before. The insertion of spill code for the 27 new local variables holding CSEs leads to the observed results. In contrast, advanced code hoisting leads to reductions of data fetches by 28.4% for the QSDPCM benchmark. The overall performance of the L1 data cache is degraded for both benchmarks (see column L1 D-Miss). For QSDPCM, an increase of cache misses by 20.7% was measured, whereas a growth from 161,422 cache misses up to 1,847,646 was observed for CAVITY.

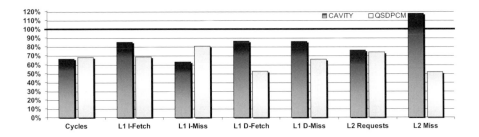

Figure 6.20. Relative Pipeline and Cache Behavior for MIPS R10000

Due to the worse data locality of CAVITY after advanced code hoisting, the amount of instruction pipeline stalls increases by 55.6% after the optimization. For QSDPCM, a reduction of stalls by 10.7% was measured. Although instruction pipeline and data cache are affected adversely by advanced code hoisting, this optimization leads to huge speed-ups for the Sun UltraSPARC III processor. The actual runtime savings are due to the fact that advanced code hoisting eliminates dozens of modulo computations and integer divisions. These operations are extremely costly for the Sun CPU so that the elimination of a modulo / division still leads to speed-ups even if the new local variable needs to be spilled to and from the main memory.

Advanced code hoisting leads to an increase of L2 cache accesses by 25.4% for the CAVITY benchmark, whereas this parameter remains unchanged in the case of QSDPCM (see column L2 Requests). With respect to L2 cache misses, a slight decrease between 2% (CAVITY) and 4.8% (QSDPCM) was observed.

This section demonstrates that the overall performance of advanced code hoisting is independent of the actual processor architecture. Despite the fact that data locality is worse after this optimization, considerable speed-ups were achieved which clearly show that advanced code hoisting is worthwhile being applied in a processor independent way. Nevertheless, it is commonly known that the size of the register file and the costs of the eliminated arithmetic operations are extremely important for a common subexpression elimination [Muchnick, 1997]. Hence, the application of advanced code hoisting under consideration of these parameters probably leads to even larger improvements as the ones shown here, but the development of such extensions towards a processor specific optimization is beyond the scope of this book.

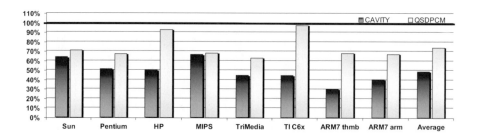

Figure 6.21.   Relative Runtimes after Advanced Code Hoisting

MIPS R10000

As can be seen from figure 6.20, advanced code hoisting applied to the CAV-
ITY and QSDPCM benchmarks leads to improvements in nearly all studied
categories in the case of the MIPS R10000.

Overall, the benchmarks are accelerated by 31.7% (QSDPCM) and 33.4%
(CAVITY). In contrast to the Sun processor, all first level caches of the MIPS
CPU benefit from the optimization. On the one hand, instruction cache fetches
go down from 14.2% (CAVITY) up to 31.5% (QSDPCM) coming along with
a simultaneous reduction of I-cache misses between 19.3% (QSDPCM) and
36.9% (CAVITY).

On the other hand, data fetches of the MIPS R10000 drop between 13.6%
(CAVITY) and 47.9% (QSDPCM) as can be seen from column L1 D-Fetch. For
this processor, the number of L1 data cache misses is also reduced. Here, a
reduction of 14.1% (CAVITY) to 34.5% (QSDPCM) was observed.

With respect to the second level unified cache, a total diminution of cache
accesses by 23.9% (CAVITY) to 26.1% (QSDPCM) can be reported here. However, a reduction of L2 cache misses was observed only in the case of the QS-
DPCM benchmark. Here, savings of 48.6% are achieved by advanced code
hoisting. In the case of CAVITY, cache misses increase by 17.7%.

6.4.2   Execution Times and Code Sizes

Diagram 6.21 clearly shows that all other processors also benefit from advanced code hoisting. In the case of the CAVITY benchmark, overall speed-ups
ranging from 33.4% (MIPS) up to 69.6% (ARM thumb) were measured. On
average over all eight processors, the source code optimization presented in this
chapter leads to a high gain of 50.8%. Compared to CAVITY, the performance
of the QSDPCM benchmark did not improve that much. Here, a maximum
acceleration of 36.9% was observed on the TriMedia architecture. In contrast,

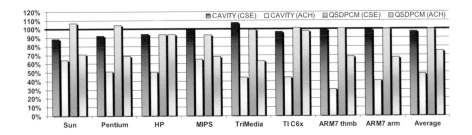

Figure 6.22.   Runtime Comparison of Advanced Code Hoisting and Common Subexpression
Elimination

advanced code hoisting only leads to a marginal speed-up of 2.3% for the TI C6x
DSP. Similarly, the HP-PA processor honors the optimization with only 6.9%
of improvement. However, an average acceleration of 25.5% was measured for
QSDPCM for all architectures listed in figure 6.21.

   In order to highlight the benefits achieved by the computation of execution
frequencies during advanced code hoisting, a direct comparison of the runtimes
after a conventional common subexpression elimination and after advanced
code hoisting is depicted in figure 6.22. This diagram clearly shows that the
application of a common subexpression elimination to the source codes of
both benchmarks does not have a significant impact on the runtimes. In some
situations, the common subexpression elimination leads to speed-ups of up
to 11.3% (CAVITY and Sun), whereas runtime degradations of up to 7.3%
(CAVITY and TriMedia) were measured in other situations. On average for
all processors considered in figure 6.22, a slight acceleration of CAVITY by
2.8% was observed after common subexpression elimination. For the QSDPCM
application having passed this optimization, the average execution times remain
unchanged. Figure 6.22 leads to the conclusion that the improvements achieved
by advanced code hoisting clearly originate from the novel steering criterion
based on the global computation of execution frequencies which is used to
control common subexpression elimination and loop-invariant code motion.

   In contrast to loop nest splitting presented in the previous chapter 5, the
high gains of advanced code hoisting do not imply an inherent increase in code
sizes. This is due to the fact that the elimination and reuse of frequently used
common subexpressions explicitly removes code from an embedded data flow
dominated application. Implicitly, some code can be added by the compiler
since advanced code hoisting augments register pressure and might lead to the
generation of additional spill code.

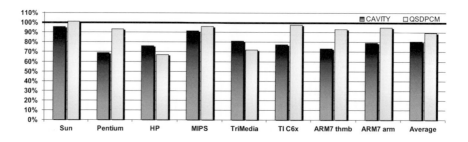

Figure 6.23.   Relative Code Sizes after Advanced Code Hoisting

Figure 6.23 shows that no code size penalties are implied by advanced code hoisting. In most cases, significant reductions of code sizes were achieved by the optimization. For the CAVITY benchmark, these reductions range from 4.4% for the Sun processor up to 31.2% for the Intel Pentium processor. In the case of the QSDPCM application, the code sizes remain unchanged by the source code optimization for the Sun and TI C6x architectures. For all other processors, improvements between 3.9% (MIPS) and 33% (HP) were obtained. The average code size reductions after advanced code hoisting for all processors amount to 10.5% for QSDPCM and 19.6% for the CAVITY benchmark.

6.4.3   Energy Consumption

This section presents experimental results showing the influence of advanced code hoisting on the energy dissipation of the studied benchmark applications for an ARM7TDMI processor. The first four columns of figure 6.24 depict the relative number of various kinds of memory accesses, whereas the last three columns show the changes in energy consumption. As usual, the 100% base line denotes the values of the unoptimized benchmarks, and the bars of the chart represent the benchmarks after advanced code hoisting in relation to the base line.

Column Instr Read of figure 6.24 clearly shows that advanced code hoisting is able to reduce the number of fetched instructions significantly. In the case of the CAVITY benchmark, a reduction of instruction fetches of 78.5% was measured, whereas this number is decreased by 33.1% for QSDPCM. Due to the elimination and hoisting of common subexpressions, the number of read accesses to the data memory is also reduced for the ARM7 core. Column Data Read shows improvements between 35.2% (CAVITY) and 33.3% (QSDPCM). The large reductions of writing memory accesses between 59.9% (CAVITY) and 26.4% (QSDPCM) – cf. column Data Write – originate from a better register allocation after advanced code hoisting for the considered platform. These

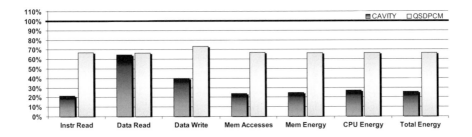

Figure 6.24. Relative Energy Consumption after Advanced Code Hoisting

detailed results lead to an overall diminution of memory accesses between 76.4% (CAVITY) and 32.8% (QSDPCM).

The savings in terms of energy consumption of the memory are of the same order of magnitude as the total reductions of memory accesses (see column Mem Energy). As can be seen, the energy consumption of the memory drops between 75.1% (CAVITY) and 33.4% (QSDPCM) after the source code optimization. For the ARM7TDMI processor, reductions of energy consumption between 72.8% (CAVITY) and 33.6% (QSDPCM) were measured. These factors lead to total energy savings from 74.5% (CAVITY) to 33.4% (QSDPCM) for the combination of memory and processor, as can be seen from column Total Energy.

6.5   Summary

This chapter presented a new source code optimization called advanced code hoisting. This technique is an elaborate combination of already known optimizations (common subexpression elimination and loop-invariant code motion) with a formal mathematical criterion steering the application of the mentioned optimizations.

It turned out that the elimination of common subexpressions might not be beneficial if the control flow surrounding an expression is not considered. As a consequence, advanced code hoisting is based on a formal model in order to compute the execution frequency of an expression. The control flow given by nested *for*-loops and *if*-statements is represented by a polytope model which is then used to compute how many times an expression surrounded by all these loops and conditions is executed.

The experimental results provided in this chapter demonstrate that this new optimization is highly beneficial. First, significant speed-ups of the studied benchmark applications were achieved. These accelerations are mainly due to the reduced number of instruction executions because of the high reuse of already computed results. Second, the fact that complex expressions are eliminated has the effect that the code sizes of the benchmarks decrease after advanced code hoisting. Third, considerable energy savings were measured which are due to the reduced amounts of accesses to main memory for fetching instructions and data.

However, it has turned out that the application of advanced code hoisting without consideration of the processor architecture might not be advantageous in some situations. For one actual processor, advanced code hoisting leads to a significant degradation of the data cache performance. Nevertheless, large speed-ups were achieved due to the savings of instruction executions. But these observations lead to the conclusion that a more sophisticated mechanism taking the register pressure and spill code generation into account can probably achieve higher gains due to a better data cache behavior. The development of strategies for adapting advanced code hoisting in a processor-dependent way at the source code level is beyond the scope of this book and is thus part of the future work.

Since advanced code hoisting is based on standard optimizations included in every current optimizing compiler, one might expect that this source code optimization does not lead to significant improvements since similar transformations would already be performed by the compilers. The fact that average speed-ups of up to 50% are achieved clearly shows that today's compilers apply their built-in optimizations only in a very restricted and limited fashion as claimed in the introduction of chapter 1. As a conclusion, it can be stated that it is worthwhile to study already known optimizations at the source code level since they are still able to outperform an optimizing compiler by far.

Chapter 7

RING BUFFER REPLACEMENT

This chapter describes two novel high-level data and control flow transformations for performance improvement of typical address-dominated multimedia applications. These transformations focus on the elimination of small arrays serving as buffers for temporary data. These arrays are generated during some data layout transformations of the already mentioned DTSE framework. They have the particular property that they are accessed in a circular way since these arrays implement ring buffer data structures.

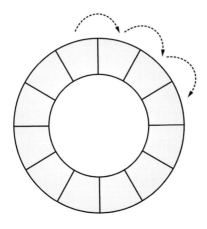

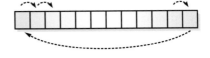

Figure 7.1.   Structure of a Ring Buffer

Figure 7.2.   Realization of a Ring Buffer using an Array

The circular access pattern to ring buffers is depicted in figure 7.1: the data is organized in such a way that the buffer does not have a dedicated first or last element. As a consequence, every element of a ring buffer has a left and right neighbor. Usually, ring buffers are implemented using arrays as illustrated

Figure 7.3.   Original Structure of the CAVITY Application

in figure 7.2. Since the inner elements of an array each have a left and right neighbor, special care only needs to be taken when accessing the left neighbor of the array's first element or the right neighbor of the last element, respectively. In such a situation, an automatic wrap-around is performed so that the last and first element of the array is chosen, resp.

In this chapter, it is shown that the elimination of these small ring buffers is advantageous and leads to considerable speed-ups and reductions of energy dissipation at the cost of only limited overhead in code size. The motivation why it is necessary to deal with ring buffer arrays is given in section 7.1. The two transformations which are executed sequentially so as to remove ring buffers completely are presented in section 7.2. A detailed presentation of experimental results is provided in section 7.3, and section 7.4 gives a short summary of this chapter.

7.1   Motivation

As already explained in previous chapters, typical multimedia applications often involve a large amount of transfers to the memories of an embedded system. This results in huge penalties with respect to runtime due to slow memories and bus systems as well as power consumption due to the domination of total system power by memory related power consumption [Wuytack et al., 1996]. In order to overcome these drawbacks, the DTSE framework of source code optimizations has previously been established.

An overview of the different tasks of DTSE is already provided in chapter 6.2 (see page 127 et sqq.) so that the description of all its individual components can be omitted here. In the following, only the steps called "data reuse" [Wuytack et al., 1998] and "in-place mapping" [Greef et al., 1997] are highlighted since the transformations of these stages are responsible for the insertion of ring buffer arrays.

The generation of ring buffer arrays during DTSE is mainly due to two reasons. First, it is attempted to make use of a predefined memory hierarchy efficiently during the data reuse step. This is done by copying frequently accessed parts of larger data into smaller arrays which can then be kept in a smaller and thus less energy consuming memory. Hereafter, in-place mapping techniques applied to these arrays (which are also called *data reuse copies*) focus at minimizing the required amount of memory by reducing the size of these data reuse copies.

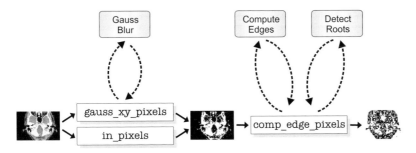

Figure 7.4. Structure of the CAVITY Application after Data Reuse Transformation

The effect of the data reuse transformation on the already described CAVITY application [Bister et al., 1989] is illustrated schematically in figures 7.3 and 7.4. In its original structure, a tomography image is processed by a pipeline consisting of three sequential steps called GaussBlur, ComputeEdges and DetectRoots. Each of these steps processes an image given as input and writes a resulting image as output which is then passed to the next stage in the pipeline.

All these steps of CAVITY process their corresponding input image line by line. The exploration of the data reuse potential of CAVITY during DTSE has shown that it is worthwhile to insert three data reuse copies which are called gauss_xy_pixels, comp_edge_pixels and in_pixels. All these data reuse copies represent buffers for entire lines of an image. gauss_xy_pixels and comp_edge_pixels contain three individual lines of the image, namely the ones currently processed by the GaussBlur and the ComputeEdges steps respectively, together with their direct neighbors [Danckaert et al., 1999, Catthoor et al., 2002]. The third line buffer in_pixels consists of only one line which is also required during the GaussBlur computation.

Immediately before the GaussBlur computations, the gauss_xy_pixels and in_pixels buffers are loaded from the input image. GaussBlur operates on these line buffers and finally flushes their contents back to the image. The same actions take place for ComputeEdges and DetectRoots and the comp_edge_pixels buffer. The structure of the CAVITY application after the data reuse transformation is depicted in figure 7.4.

These line buffers are subject to the subsequent in-place mapping stage of DTSE. During this optimization, a formal model is constructed representing the accesses to individual elements of the line buffers across all points of the

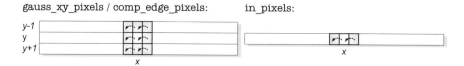

Figure 7.5.   Access Analysis of Line Buffers during In-Place Mapping

ure 7.5. As a consequence, it is not necessary to store the lines of the image in their entirety in the buffers. Instead, only the neighboring pixels of the current (x, y)-position need to be saved. This way, the line buffers gauss_xy_pixels and comp_edge_pixels can be reduced to two very small two-dimensional arrays of size 3×3. in_pixels has a size of 3 after in-place mapping.

During the transition from position (x, y) to $(x + 1, y)$, the second and third column of these smaller line buffers would have to be copied to the first and second column respectively as indicated by the arrows in figure 7.5. Additionally, the third column needs to be filled with new pixels of the image. In order to avoid these copy operations, the line buffers are accessed in a ring buffer fashion. This is implemented by computing the index expression for the line buffer modulo 3.

EXAMPLE 7.1

For the position x $= 769$, *an access to the line buffer* in_pixels *like*

   in_pixels[x % 3] = ...;

overwrites the middle element of in_pixels, *since* $769\ \%\ 3 = 1$. *Analogously, the expressions* in_pixels[(x-1) % 3] *and* in_pixels[(x+1) % 3] *access the leftmost and rightmost elements respectively.*

After the transition from x *to* x $+ 1$, x *is equal to* 770. *For* in_pixels[x % 3], *this position now references the right column of the line buffer since* $770\ \%\ 3 = 2$. *Compared to the situation where* x $= 769$, *it can be observed that the line buffer is now accessed one position further to the right which is exactly the desired behavior.*

It is easy to see that the data reuse and in-place mapping steps of DTSE are highly efficient in exploring the locality of data accesses. Array references causing most of the main memory accesses and the inherent energy dissipation are identified and eliminated. After the transformation, most of the computations are performed using very small arrays containing temporary data which can be kept in small and energy-efficient memories that are close to the processor.

Despite this closeness of data to the processor, the storage of the line buffers in the processor's registers – which are by far the fastest and least energy consuming memories – is virtually disabled. This is due to the fact that data reuse copies are organized in arrays. Only very few compilers even try to allocate subscripted variables to registers, despite the fact that register allocation is often done very efficiently for scalars [Muchnick, 1997].

```
for (i=0; i<n; i++)                    for (i=0; i<n; i++)
   for (j=0; j<n; j++)                    for (j=0; j<n; j++) {
                                             c = C[i][j];
                            →
      for (k=0; k<n; k++)                     for (k=0; k<n; k++)
         C[i][j] += A[i][k]*B[k][j];             c += A[i][k]*B[k][j];

                                             C[i][j] = c; }
```

Figure 7.6. Scalar Replacement of Array Elements

Scalar replacement of array elements [Muchnick, 1997] tries to store frequently accessed array elements in scalar variables which are subject to register allocation (see figure 7.6). This optimization requires a careful analysis of the index expressions used to reference an array, which is generally unable to deal with complex expressions containing modulo operations generated during DTSE. Hence, scalar replacement of array elements is not applied to data reuse copies in practice.

Another major disadvantage of the steps of DTSE described here is the generation of very complex addressing code. The instruction sets of various processors do not support modulo computations. Instead, these operations are provided by runtime libraries, so that the processor has to perform a costly call to the library. A software realization of the modulo operator by the runtime library requires an enormous amount of clock cycles. These properties make array accesses using modulo computations a performance bottleneck for a DTSE optimized embedded data flow dominated application. In [Ghez et al., 2000], techniques for reducing the overhead of modulo addressing were presented. However, the virtually disabled storage of data reuse copies in processor registers is not addressed so that this publication is considered to be complementary to the ring buffer replacement techniques presented here.

Numerous modern embedded processors contain dedicated *address generation units* (AGUs) that compute addresses for memory accesses in parallel to the data paths. The AGUs of several processors (e. g. Motorola DSP56600 [Motorola Inc., 1996], TI C6x [Texas Instruments Inc., 1999], Infineon TriCore [Infineon Technologies, 2001]) already support the so called *modulo addressing* (also known as *circular addressing*). This addressing mode is explicitly intended to access ring buffers in a circular way. Modulo addressing seems to be perfectly suited for the execution of code generated by the DTSE transformations. However, the results given in this chapter imply that even the Texas Instruments compiler is unable to detect from the C source code that the line buffers of the CAVITY application are organized in a circular way. A study of the assembly code emitted by this compiler has shown that modulo addressing is definitely not used by the compiler so that the programmer would have to use assembly instructions in order to use these advanced features of the TI's AGU.

This chapter presents a non-algorithmical approach which is able to solve all problems mentioned above. Since compilers are often unable to analyze the behavior of DTSE optimized source codes accurately, it is the goal of the *ring buffer replacement* techniques to make the properties of the data reuse copies explicit to the compiler. In a very abstract way, this is done as depicted in figure 7.7.

```
int in_pixels[3];                        int ip0, ip1, ip2;

for (x=0; x<n; x++) {                     for (x=0; x<n; x++) {
  in_pixels[x % 3] = img[...];             ipₓ = img[...];
  a = in_pixels[(x-2) % 3] * 68;    →      a = ip_y * 68;
  a += in_pixels[(x-1) % 3] * 99;          a += ip_z * 99;
  a += in_pixels[x % 3] * 68;              a += ipₓ * 68;
  ... }                                    ... }
```

Figure 7.7.   Principal operating Mode of Ring Buffer Replacement

The left hand side of figure 7.7 represents the x-loop of the CAVITY application scanning one line of an image after DTSE. It can be seen that one pixel of the line buffer in_pixels at the position x % 3 is initialized with a pixel of the tomography image first. After that, some intermediate value a is computed using the actual contents of the buffer in_pixels. It is important to see that in_pixels is organized as an array of three elements as described previously.

On the one hand, ring buffer replacement performs a scalarization of the array in_pixels. This can be seen on the right hand side of figure 7.7 where in_pixels is replaced by three new integer variables denoted ip0, ip1 and ip2. On the other hand, this scalarization is performed in such a way that the cyclic accesses to the original ring buffer are modeled correctly using the new scalar variables. This property is illustrated in an abstract way in the above figure. In the body of the x-loop, all accesses to the ring buffer in_pixels are replaced by accesses to the scalar variables. These variable accesses are shown in the figure using abstract subscripts x, y and z for the sake of simplicity. It can be seen that the first and fourth line of the loop body both access the same scalar variable. This corresponds to the original behavior since the same lines on the left hand side of figure 7.7 both access the array element in_pixels[x % 3].

The use of the subscripts ip_x, ip_y and ip_z in the figure is chosen here in order to hide the technical details how the circular access pattern to the scalarized ring buffer is actually modeled. The generic subscripts simply illustrate the fact that during each iteration of the x-loop, a different scalar variable is accessed in every line of the loop body. The way how this behavior is implemented is described in the following section 7.2.

The advantages of ring buffer replacement as illustrated in figure 7.7 are manifold:

- Address computations are completely removed from the source code. In particular, the expensive modulo operations are eliminated, leading to significant speed-ups.

- The replacement of data reuse copies by local variables automatically makes the scalarized arrays a target for the register allocator of a compiler. Hence, ring buffer replacement explicitly enables the possible storage of data reuse copies in the register file of a processor.

- In the case that the newly generated local variables can not be kept in registers, the scalarization of a ring buffer has the additional effect that the memory addresses of the individual scalar variables are determined statically at compile time. As a consequence, dynamic address computations at least consisting of the addition of a base address with an offset are entirely eliminated. This makes even DSPs equipped with sophisticated address generation units benefit.

- It is well-known that compiler optimizations like e. g. constant folding and propagation, copy propagation or dead code elimination mainly focus on scalar variables instead of arrays since a precise array dependence and alias analysis is a much harder task compared to a scalar data flow analysis. Thus, ring buffer replacement explicitly enables the application of these compiler optimizations to the data reuse copies generated during the DTSE transformations.

These benefits of ring buffer replacement come along with an increase in code size which is shown to be moderate and which can be avoided under some circumstances, depending on actual processor architectures.

7.2   Optimization Steps

In order to perform a replacement of ring buffer arrays as sketched in figure 7.7, a two step approach is described in this section. First, the scalarization of the data reuse copies taking the circular access pattern into account is required (see section 7.2.1). After that, a loop unrolling steered by parameters extracted from the ring buffers is performed (see section 7.2.2) so as to eliminate some overhead inserted by the scalarization step.

7.2.1   Ring Buffer Scalarization

In the first phase of ring buffer replacement, the elements of a circular data reuse copy are scalarized such that explicit addressing code like for example (x-2) % 3 (see left hand side of figure 7.8) and the implicit addition with the base address of the array are removed from the code.

```
int in_pixels[3];                          int ip0, ip1, ip2;

for (x=0; x<n; x++) {                      for (x=0; x<n; x++) {
                                               ip2 = ip1;
                                               ip1 = ip0;
    in_pixels[x % 3] = img[...];      →        ip0 = img[...];
    a = in_pixels[(x-2) % 3] * 68;             a = ip2 * 68;
    a += in_pixels[(x-1) % 3] * 99;            a += ip1 * 99;
    a += in_pixels[x % 3] * 68;                a += ip0 * 68;
    ... }                                      ... }
```

Figure 7.8.   Ring Buffer Scalarization

After replacing the array with a set of scalar variables, a one-to-one mapping of the individual array elements to the new variables has to be defined. In the example shown in figure 7.8, the actual buffer element in_pixels[x % 3] is mapped to variable ip0. Likewise, the preceding elements in_pixels[(x-1) % 3] and in_pixels[(x-2) % 3] are represented by ip1 and ip2, respectively, after the scalarization step. In the loop body, every access to the ring buffer array has to be replaced by variable accesses according to the defined mapping. This is illustrated on the right hand side of figure 7.8.

Special care has to be taken to model the circular access to the elements of the ring buffer correctly. This is done by inserting some copy instructions at the beginning of a loop as shown in lines three and four of the right hand side of figure 7.8. Using these copy instructions, the contents of the scalar variables are shifted by one position during every loop iteration. The data stored in ip2 before the execution of the copy instructions is unneeded during further iterations of the x-loop. As a consequence, ip2 and ip1 can be overwritten by the copy instructions as depicted in figure 7.8. After shifting, the data originally stored in ip0 resides in ip1 so that ip0 can legally be overloaded with a new pixel of the image. This process is also illustrated in the following example 7.2.

EXAMPLE 7.2

This example shows the correspondence between the original ring buffer and its scalarized counterpart by means of the source codes shown in figure 7.8.

For the iteration x = 769, *the expression* `in_pixels[x % 3]` *accesses array element number 1. In iteration* x = 770, *the same array element is accessed by the expression* `in_pixels[(x-1) % 3]`.

According to the mapping of array elements to scalar variables described above, variable `ip0` *refers to* `in_pixels[x % 3]`. *The mapping of array elements to variables also defines that element* `in_pixels[(x-1) % 3]` *is represented by* `ip1`.

For the code in the right part of figure 7.8 to be correct, it is has to be ensured that the data stored in `ip0` *during iteration* x = 769 *is stored in* `ip1` *for the next value* x = 770. *This is the task of the copy instruction* `ip1 = ip0` *at the beginning of the loop body. It obviously has the effect that the access to* `ip1` *in the current iteration references the same data that was accessed via* `ip0` *during the preceding iteration. It is simple to see that the same holds for* `ip2`. *Hence, the copy instructions at the beginning of the loop correctly model the circular structure of a ring buffer.*

7.2.2 Loop Unrolling for Ring Buffers

The scalarization step described in the previous section leads to a slight overhead, because additional instructions and data transfers are inserted. This kind of overhead can not be removed using standard compiler optimizations like e. g. copy propagation. In order to overcome this problem, a new criterion for the commonly known loop unrolling optimization is presented in the following.

Conventional loop unrolling (see also page 22) as described in [Muchnick, 1997] is a very common technique for exploring instruction-level parallelism and reducing loop overhead. The determination of the so-called *unrolling factor* is normally based on an analysis of the size and the number of iterations of a loop in order to avoid explosions of code sizes.

In contrast, loop unrolling in the context of ring buffer replacement is steered by the number of copy instructions inserted by the scalarization step described in section 7.2.1. The left hand side of figure 7.9 shows three variables created during ring buffer scalarization. The contents of variables `ip1` and `ip0` are copied to `ip2` and `ip1` respectively, and a new value is assigned to `ip0`. These two copy instructions can be removed completely by unrolling the x-loop with a factor of three.

EXAMPLE 7.3

For the source code still containing the original ring buffer `in_pixels` *(see left hand side of figure 7.8), three different combinations of accesses to the original ring buffer exist:*

- `in_pixels[x % 3]` *refers to element 0 of the array,* `in_pixels[(x-1) % 3]` *to element 2, and* `in_pixels[(x-2) % 3]` *to element 1.*

- `in_pixels[x % 3]` *refers to element 1 of the array,* `in_pixels[(x-1) % 3]` *to element 0, and* `in_pixels[(x-2) % 3]` *to element 2.*

- `in_pixels[x % 3]` *refers to element 2 of the array,* `in_pixels[(x-1) % 3]` *to element 1, and* `in_pixels[(x-2) % 3]` *to element 0.*

```
                              int ip0, ip1, ip2;

                              for (x=0; x<n; x+=3) {
                                  ip2 = img[...];      /* Copy 1 */
                                  a = ip1 * 68;
int ip0, ip1, ip2;                a += ip0 * 99;
                                  a += ip2 * 68;
for (x=0; x<n; x++) {             ...
    ip2 = ip1;
    ip1 = ip0;                    ip1 = img[...];      /* Copy 2 */
    ip0 = img[...];    →          a = ip0 * 68;
    a = ip2 * 68;                 a += ip2 * 99;
    a += ip1 * 99;                a += ip1 * 68;
    a += ip0 * 68;                ...
    ... }
                                  ip0 = img[...];      /* Copy 3 */
                                  a = ip2 * 68;
                                  a += ip1 * 99;
                                  a += ip0 * 68;
                                  ... }
```

Figure 7.9.   Loop Unrolling for Ring Buffers

By using the unrolling factor of three, every copy of the unrolled loop can exclusively treat one of the above combinations.

This loop unrolling has the effect that every possible combination of accesses to the original ring buffer is explicitly handled by its own part of the unrolled loop. By adapting the definitions and uses of the scalarized variables appropriately for every copy of the unrolled loop, the copy instructions created during the scalarization step become redundant and can be removed from the code. The outcome of the loop unrolling phase for ring buffers is depicted on the right hand side of figure 7.9.

It is easy to see that the code structure resulting from this loop unrolling step meets the goals formulated in section 7.1. On the one hand, all address computations related to the original ring buffer array are eliminated entirely. O

7.3   Experimental Results

In contrast to the techniques presented in the previous chapters, ring buffer scalarization and loop unrolling for ring buffers have not been automated. Since no formal models for the implementation of these source code optimization techniques are developed in this book, both transformations were applied manually to the CAVITY application which serves as an example throughout this chapter. The figures presented in this section represent the behavior of this benchmark after each of the scalarization and loop unrolling steps. Like in the previous chapters, the 100% base lines represent the values for the original unoptimized version of CAVITY.

In analogy to the structure of the preceding chapters, the influence of ring buffer replacement on instruction pipelines and caches is shown in section 7.3.1. The speed-ups and the corresponding enlargements of CAVITY are subject of section 7.3.2, whereas the changes in terms of energy dissipation are presented in section 7.3.3. Appendix D lists the data measured during the benchmarking of ring buffer replacement in more detail.

7.3.1   Pipeline and Cache Performance

Intel Pentium III

Figure 7.10 depicts the behavior of the CAVITY application after both the ring buffer scalarization and the loop unrolling steps. As can be seen from column Cycles, ring buffer replacement as proposed in this chapter is able to speed up the CAVITY benchmark by 20.9%. This overall acceleration is exclusively caused by the scalarization step which leads to a decrease of runtimes of 19.9%. Loop unrolling only contributes with one percent to the measured speed-up. The optimizations only have a marginal effect on the instruction pipeline. After loop unrolling, a decrease of taken branches by 2.7% was observed which is due to the reduced loop overhead after unrolling.

It it obvious that the acceleration of the benchmark is due to the reduction of executed instructions (see column L1 I-Fetch). The total amount of executed instructions drops by exactly the same order of magnitude as the execution times. The fact that the addressing overhead for three line buffers with a total size of only 21 elements contributes 20% to the total runtime of the benchmark clearly shows that it is necessary to optimize the addressing code of DTSE transformed applications. A scalarization is the most powerful possible addressing optimization since it removes address code entirely. This improvement has a positive effect on the behavior of the first level instruction cache, since the number of I-cache misses is reduced by 13.2% after ring buffer replacement. After the scalarization step only, an even higher improvement of 20.4% was measured. The slight penalty caused by loop unrolling is obvious since this transformation enlarges the loop body leading to the modified cache behavior.

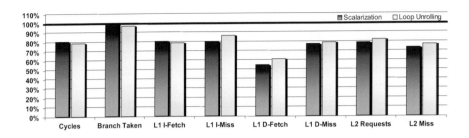

Figure 7.10.   Relative Pipeline and Cache Behavior for Intel Pentium III

It is remarkable that the total number of data cache accesses is reduced heavily. After ring buffer scalarization, a reduction of 45.2% was observed. This result clearly underlines that the focus of the optimizations on small arrays storing temporary data is justified. The assumption that ring buffer scalarization enables the optimization potential of the compiler to be applied to the former arrays is approved. These large reductions of data memory accesses are only possible if the compiler is able to efficiently store the data in the processor's registers. The application of loop unrolling leads to an increase of data fetches by 6.3% which can be explained with a worse register allocation for the large loop body. The diminutions of D-cache accesses explained so far lead to a reduction in D-cache misses between 22.4% after scalarization and 20.6% after loop unrolling.

Since the number of misses for both first level caches drops by roughly 20%, the number of accesses to the second level unified cache of the Intel Pentium is reduced by the same order of magnitude. Actually, reductions of 20.9% (scalarization) and 17.5% (loop unrolling) were measured (see column L2 Requests). As a consequence, data and instruction transfers involving main memory go down by 26.3% (scalarization) and 22.7% (loop unrolling) (cf. column L2 Miss).

Sun UltraSPARC III

In the case of the Sun UltraSPARC III processor, even larger improvements were observed compared to the Intel Pentium (see figure 7.11). As can be seen from column Cycles, accelerations of 18.4% (scalarization) and 21.5% (unrolling) were observed. For this processor, the loop unrolling step implies visible changes. Column Branch Taken shows that this step leads to an improvement of the branching behavior of 5.9% due to the reduced control flow overhead afterwards.

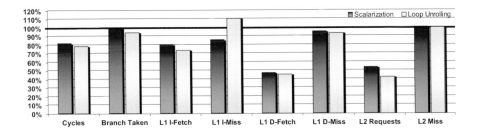

Figure 7.11. Relative Pipeline and Cache Behavior for Sun UltraSPARC III

The overhead caused by the complex addressing of the ring buffers amounts to 19.9% (see the bar of column L1 I-Fetch representing the scalarization step). A further reduction of instruction executions up to 26.3% is achieved by loop unrolling. It can be concluded that the copy instructions inserted during the ring buffer scalarization constitute an overhead of approximately 5% for the Sun processor which is eliminated by loop unrolling. The influence of the individual steps of transformations on the amount of I-cache misses is comparable to the behavior of the Intel processor. Column L1 I-Miss of figure 7.11 shows a gain of 14.5% after scalarization. Unlike the Pentium processor, loop unrolling leads to a total increase of cache misses by 10.8% in the case of the Sun UltraSPARC.

The application of ring buffer replacement is extremely beneficial for the L1 data cache of the Sun processor. As can be seen from column L1 D-Fetch, reductions of cache accesses between 52.7% after scalarization and 54.4% after loop unrolling are achieved. Again, these results demonstrate the effectiveness of the approach to keep frequently used data as close to the functional units as possible. Due to the huge size of the data cache, the reductions of cache misses are not as large. Here, savings up to 6.1% after loop unrolling were measured.

The number of accesses to the L2 cache is reduced by up to 57.3% after loop unrolling. The significance of this transformation can be seen from column L2 Requests, since it makes up a difference of more than 10% compared to ring buffer scalarization. These benefits do not have an influence on the number of L2 cache misses – their number remains constant during all of the transformations.

MIPS R10000

Column Cycles of figure 7.12 shows an overall acceleration of the CAVITY application by 32%. Ring buffer scalarization alone already leads to a speed-up of 31.2%. The unrolling of the x-loop with a factor of three leads to a reduction of executed branch instructions by 50.3%.

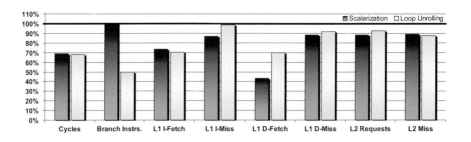

Figure 7.12.   Relative Pipeline and Cache Behavior for MIPS R10000

The removal of addressing instructions achieved by the scalarization of the arrays has the effect that 26.6% less instructions are fetched by the MIPS R10000. The influence of the copy instructions and of the control flow overhead which are subject to the loop unrolling step amount to another 3% leading to total savings of 29.6% after all optimizations. With respect to L1 instruction cache misses, savings of 13.4% were measured. After the subsequent loop unrolling, only a marginal reduction of 1.2% remains (see column L1 I-Miss).

The gains of ring buffer scalarization concerning data fetches from the L1 D-cache are even higher than for the Intel and Sun processors. In the case of the MIPS R10000, a reduction of D-fetches by 56.4% was measured. Due to the register allocator of the MIPSpro compiler, an increase occurs after loop unrolling. Nevertheless, a total reduction of 29.9% was achieved after the application of both transformations. The influence of these optimizations on the first level data cache misses lies around 10%: reductions between 11.8% (scalarization) and 8.3% (loop unrolling) were observed.

Fetches and misses of the L2 cache are likewise reduced. Column L2 Requests of figure 7.12 shows a diminution of L2 cache accesses by 12% after scalarization (7.4% after loop unrolling). The number of cache misses drops between 11.1% (scalarization) and 12.2% (loop unrolling) which is illustrated by column L2 Miss.

7.3.2   Execution Times and Code Sizes

Figure 7.13 shows that the speed-ups measured for the Sun, Intel and MIPS processors also occur when various other processors are considered. As can be seen, the scalarization step is the most effective part of ring buffer replacement. The accelerations due to this optimization vary between 9.7% for the TriMedia and up to 31.2% for the MIPS processor. Only in the case of the HP processor, no effect of the scalarization was observed. On average for all processors considered in figure 7.13, the execution times of CAVITY decrease by 17.7%.

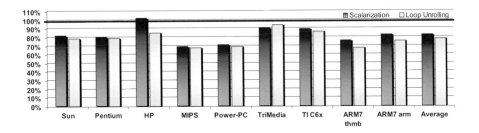

Figure 7.13. Relative Runtimes after Ring Buffer Replacement

Similar to the observations presented in the previous section 7.3.1, the influence of loop unrolling on the runtimes varies among the processors. For the Pentium, MIPS and Power-PC architectures, no significant effect of loop unrolling could be observed. However, this technique is beneficial for the Sun, HP, TI and ARM processors. Here, accelerations between 2.6% (TI C6x) and 16.7% (HP) were measured. In the case of the TriMedia processor, a degradation of performance by 4% occurred. The entire sequence of ring buffer scalarization and loop unrolling leads to an average total speed-up of 21.7% for all processors.

The effect of both steps of ring buffer replacement on code sizes is depicted in figure 7.14. As can be seen, the scalarization of data reuse copies leads to considerable reductions of code sizes. These diminutions vary between 9.11% for the TriMedia and up to 24.3% for the Sun processor. As previously demonstrated by means of the number of executed instructions, these results show the significance of addressing code in the CAVITY application.

Loop unrolling applied after the scalarization naturally leads to increases in code sizes. These augmentations are greater than the savings achieved by ring buffer scalarization so that the code sizes after loop unrolling are larger than those before the ring buffer replacement. However, figure 7.14 shows that the total code size increase due to ring buffer replacement only varies between 16.9% (Sun) and 42.1% (HP). Since the x-loop of CAVITY is unrolled by a factor of three and since almost all the code of the application resides in this loop, a code size increase between a factor of two and a factor of three would be expected after ring buffer replacement. In this context, the actually measured growths of 30% are extremely moderate.

The figures given in this section show that loop unrolling is only beneficial for some of the architectures studied here. If code size increases have to be avoided by all circumstances, a suitable approach is to quantify the effects of loop unrolling for ring buffers for an actual processor architecture. If it does not

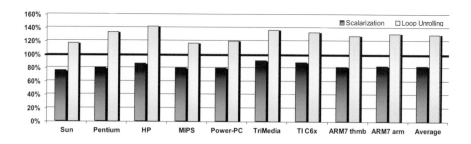

Figure 7.14. Relative Code Sizes after Ring Buffer Replacement

contribute significantly to the total acceleration of an embedded application, it is worthwhile to apply only the ring buffer scalarization step. This way, achievable speed-ups can be traded off with possible code size increases.

7.3.3 Energy Consumption

In the following, the energy savings due to ring buffer replacement are presented for an ARM7TDMI processor. In order to obtain these experimental results, ring buffer scalarization and loop unrolling for ring buffers were applied to the source code of the CAVITY application. For both resulting code versions, i. e. that after scalarization and that after scalarization and loop unrolling, energy profiling as described in section 3.3.3 (see page 37) was performed leading to the diagram shown in figure 7.15. As usual, the contents of this figure is shown relatively as a percentage of the unoptimized CAVITY application which is represented by the bold 100% line.

The impact of the addressing code executed for accessing the ring buffers of CAVITY can be seen from column Instr Read. It can be seen that ring buffer scalarization leads to a decrease of executed instructions by 20.1%. The influence of the copy instructions which are removed during loop unrolling for ring buffers is limited since loop unrolling only leads to a further decrease of executed instructions by 1.7%.

Columns Data Read and Data Write illustrate the effect of the optimizations on the total amount of memory accesses. Due to the entire removal of addressing code depending on the loops' index variables, a reduction of reading memory accesses by 24.9% was measured after ring buffer scalarization. The elimination of the inserted copy instructions leads to a further reduction up to 36.7% in total after loop unrolling. In the context of writing memory accesses, the additional improvements achieved by loop unrolling are even larger. Here, this step leads to a total reduction of memory accesses by 25.1%, whereas the benefits of the scalarization only amount to 3.8%. These numbers indicate the influence of the

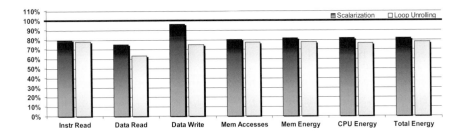

Figure 7.15. Relative Energy Consumption after Ring Buffer Replacement

copy instructions generated during ring buffer scalarization since the shifting of the contents of the new scalar variables by one position leads to those writing memory accesses which are subject to loop unrolling. In their combination, the individual steps of ring buffer replacement lead to a total reduction of memory accesses (column Mem Accesses) by 19.9% after scalarization and by 22.9% after loop unrolling.

As a consequence, the energy dissipation of the main memory is reduced by the same order of magnitude. As can be seen from column Mem Energy, the energy consumption drops by 18.3% (scalarization) and in total by 22.3% after both scalarization and loop unrolling. Similar savings were measured for the ARM7 core, where reductions by 18.3% and 23.7% were observed after scalarization and loop unrolling, respectively (see column CPU Energy). For the entire system consisting of memory and processor, total energy reductions by 18.3% after scalarization and by 22.7% after the entire sequence of optimizations were obtained.

7.4 Summary

The source code transformations presented in this chapter focus on small arrays which are accessed in a cyclic manner. These ring buffers are generated during the data reuse and in-place mapping steps of source code transformation frameworks for memory hierarchy exploration (DTSE). The techniques informally described here are able to replace these arrays by a set of scalar variables. In a first step, the scalarization of the arrays is performed, followed by a loop unrolling phase driven by the characteristics of the ring buffers. This way, the cyclic access pattern to the data stored in the former arrays is maintained in contrast to existing scalarization techniques which are unable to deal with such arrays.

The experimental results provided for an embedded data flow dominated application having passed the DTSE transformations show that it is necessary to perform optimizations on these small ring buffers. The application of ring buffer replacement leads to an average speed-up of 21.7% with a simultaneous code size increase of only 30%. These high gains mainly stem from two effects caused by ring buffer replacement:

- First, the replacement of arrays by local variables entirely removes all code required for addressing the arrays.

- Second, the newly created scalar variables are automatically considered by the register allocator of a compiler so that the contents of the ring buffers can be stored in processor registers for the first time.

It is worth mentioning that the results given in this chapter clearly underline the effectiveness of the DTSE transformations. The fact that an optimization potential of around 20% is explored by ring buffer replacement implies that the benchmark application requires around one fifth of its entire runtime for accessing the small ring buffers. This is a clear indicator of the extremely high data locality achieved by the DTSE transformations.

The experimental results show that the influence of the loop unrolling step depends on the actual processor architecture. If an average code size increase of 30% is not tolerable, it can be sufficient to apply only the ring buffer scalarization step which generally leads to high speed-ups with a simultaneous reduction of code sizes. It is part of the future work to automate the optimizations for ring buffer replacement using formal models to analyze the cyclic access patterns to arrays.

Both the ring buffer scalarization and the loop unrolling for ring buffers were briefly published in [Falk et al., 2003a].

Chapter 8

SUMMARY AND CONCLUSIONS

This book presented source code optimization techniques for data flow dominated embedded software. It represents the attempt of bridging the gap between existing high-level source code optimization frameworks and today's optimizing compilers. Section 8.1 of this chapter summarizes and evaluates the results presented in this work. After this, an overview of future work given in section 8.2 concludes this book.

8.1   Summary and Contribution to Research

Code optimization for embedded software has turned out to be an issue of high importance since the demands on embedded systems with respect to runtime and power efficiency are increasing. Traditionally, code optimization was performed at the level of compilers for embedded processors. Recently, the application of optimization techniques at the level of program source codes which are fed into optimizing compilers afterwards has attracted a high interest.

A study of existing source code transformation frameworks has shown that massive improvements are achievable. In particular, optimization frameworks focusing on the exploration of memory hierarchies have proven to be highly beneficial. Despite the considerable results of these memory hierarchy optimization frameworks, this book demonstrated that a very large potential for optimization at the source code level still is not explored at all. This is mainly due to the sole concentration on data flow optimizations leading to the situation that control flow issues are not considered at all. Actually, an even worse structure of control flow results from the data flow and memory hierarchy optimizations.

For these reasons, this book presented optimization techniques applied at the source code level which explicitly take control flow issues into account. These optimizations focus on data flow dominated embedded applications and are one

181

the one hand kept very general so that many of them can be applied to arbitrary data flow dominated source codes. On the other hand, the optimizations are intended to remove the control overhead imposed by the aforementioned memory hierarchy optimizations.

In more detail, the following optimization techniques were presented in this work:

Loop Nest Splitting is able to perform a complex control flow analysis of nested loops and *if*-statements. This way, iterations of a loop nest where all considered *if*-statements are provably satisfied can be identified. Using this information, the entire loop nest is split such that the number of executed *if*-statements is minimized.

Advanced Code Hoisting is an optimization which is based on already existing simple transformations, namely common subexpression elimination and loop-invariant code motion. These standard techniques are combined by means of a new criterion leading to the minimization of instruction executions after the optimization. In order to minimize instruction executions, execution frequencies are computed based on an accurate representation of nested loops and *if*-statements.

Ring Buffer Scalarization aims at minimizing the overhead required for addressing small arrays serving as ring buffers for temporary data. This goal is achieved by replacing such arrays by a set of scalar variables. This way, no addressing code is required at all, and the scalarized arrays are likely to be kept in the processor's registers. A slight overhead is inserted into the source codes in order to model the cyclic accesses to the former ring buffers correctly.

Loop Unrolling for Ring Buffers can be seen as a post-processing phase for the ring buffer scalarization technique. By unrolling selected loops containing accesses to scalarized ring buffers, the already mentioned overhead – which consists of some copy instructions during every loop iteration – can be removed entirely from the source code at the cost of a modest increase in code size.

A summary of the effect of these optimizations on the runtime of a realistic benchmark application (CAVITY) for a large series of different processors is depicted in figure 8.1. This diagram shows the runtimes of the benchmark for each phase of the entire sequence of optimization. The optimizations were applied in the following order:

1 Advanced Code Hoisting

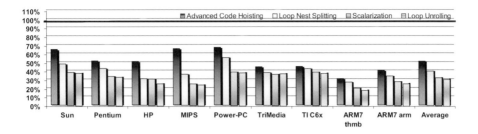

Figure 8.1. Relative Runtimes after entire Optimization Sequence

2 Loop Nest Splitting[1]

3 Ring Buffer Scalarization

4 Loop Unrolling for Ring Buffers

In order to enable the easy comparison of the various results, the runtimes of figure 8.1 are given relatively as a percentage of the original unoptimized source code version which is highlighted by the 100% base line.

As can be seen from figure 8.1, considerable speed-ups are achieved by the optimization sequence for all studied processors. These total gains vary between 61.5% for the Power-PC and up to 81.7% for the ARM7 in thumb mode. The average speed-up for all processors amounts to 68.8% which is more than a factor of three of improvement. Although individual optimizations contribute to a different extent to the measured improvements, these results clearly demonstrate that the optimizations presented in this book are highly independent of actual processor architectures.

These large speed-ups mainly originate from the reductions of total instruction executions achieved by all optimizations. Loop nest splitting heavily reduces executions of *if*-statements and complex conditions. Advanced code hoisting minimizes address code execution which is also targeted by the ring buffer optimizations. Since it is well-known that transfers of instructions and data between background memories and a processor contribute most to the total energy dissipation, the entire optimization sequence presented above also leads to massive energy savings. This fact is depicted in figure 8.2. As can be seen

---

[1] Since the optimization potentials explored by loop nest splitting and advanced code hoisting do not mutually interfere, the ordering in which these optimizations are applied is generally irrelevant. Due to technical reasons imposed by the SUIF intermediate representation, it was necessary to perform loop nest splitting after advanced code hoisting here.

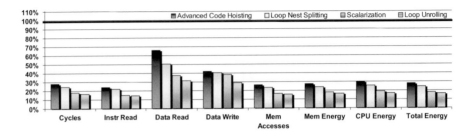

Figure 8.2.   Relative Energy Consumption after entire Optimization Sequence

from column Mem Accesses, the entire optimization sequence leads to a reduction of memory accesses by 84.4%. As a consequence, the energy consumed by the memory and the ARM processor is reduced by a similar order of magnitude. Total energy savings of 82.9% were measured after all optimizations presented in this book.

Besides the achieved savings in execution time and energy dissipation, the main contributions of the optimization techniques presented in this book are:

- The optimizations presented in this work have proven to be effective irrespective of actual processor architectures.

- For the most sophisticated techniques, algorithms are presented and fully automated tools for optimization exist.

- By combining the two previous items, it can be concluded that this book presents high-level optimizations whose scope is beyond that of actual compilers on the one hand. On the other hand, their almost full automation principally enables their integration into future compilers.

- Hence, this work successfully bridges the gap between existing high-level source code optimization frameworks and today's optimizing compilers.

8.2   Future Work

This book ends with a short outlook on the most relevant issues for future research, sorted by the individual optimization techniques:

Loop Nest Splitting

It is obvious from chapter 5 that the code size increases resulting from loop nest splitting as described in this book are a severe drawback of this optimization. However, loop nest splitting offers inherent opportunities for solving this

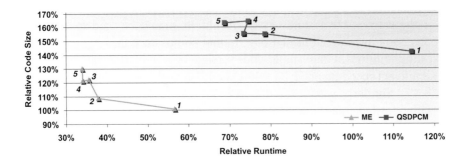

Figure 8.3. Possible Speed / Size Trade-Offs for Loop Nest Splitting

problem since it is perfectly suited for trading off speed-ups with code size increases.

During the presentation of the genetic algorithms used to perform loop nest splitting (see sections 5.3.3 and 5.3.5), the notion of the loop λ containing the splitting-if is introduced (see example 5.6 on page 75). λ is defined in such a way that it always refers to the outermost possible loop which can legally contain the splitting-if in order to achieve a maximum speed-up. This results in the transformed loop nest depicted in figure 5.3 (see page 59) where the splitting-if if (x >= 10 || y >= 14) is placed in the y-loop. Within this *if*-statement, the remaining loop nest consisting of the k-, l-, i- and j-loop can be found.

Since the splitting-if does not depend on index variables occurring inside λ by definition, it is always legal to set λ to any of these loops – e. g. λ can refer to the k-, l-, i- or j-loop for the ME example. In such a situation, the portions of the loop nest which have to be replicated in order to perform loop nest splitting obviously become smaller on the one hand. On the other hand, it can be expected that only smaller speed-ups will be achieved since more *if*-statements are executed, leading to the mentioned trade-off.

Figure 8.3 shows the pareto curves for the ME and QSDPCM benchmarks which represent this speed / size trade-off. The x-axis represents the relative runtimes of the benchmarks, whereas the y-axis shows the corresponding relative code sizes (100% $\hat{=}$ unoptimized code version). The figure was generated using manually generated source codes. Those points of the curves marked with the number "1" represent the codes where λ refers to the innermost loop of the entire loop nests. The points labeled with "5" denote the placement of λ in the outermost possible loop, i. e. the code generated by the techniques presented in chapter 5.

As expected, the "5" codes lead to the highest speed-ups, whereas version "1" is the slowest but smallest one. Attention should be paid to the code versions lying in-between these two extremal points since they represent very interesting solutions. For example, code version "2" of the ME benchmark leads to a significant speed-up of 62.1% while simultaneously increasing code size by only 8.8%. Similarly, version "3" of QSDPCM is quite fast (26.6% acceleration) and with 55.3% increase one of the smallest versions of QSDPCM.

These manual experiments clearly show that it is worthwhile to study the possible trade-offs when applying loop nest splitting under tight code size constraints. A more systematic study than the one presented in this chapter resulting in an automated approach for exploring these speed / size trade-offs is part of the future work.

Advanced Code Hoisting

Classical literature on compiler construction suggests not to perform a common subexpression elimination without considering compiler and processor internal information about register pressure and the size of the register file since this may lead to a performance degradation of a program under particular circumstances.

In the context of advanced code hoisting, an obvious runtime degradation was not observed for any of the studied processors. In every case, large speed-ups and energy savings were measured, justifying the approach presented in chapter 6 which only focuses on the structure of a program to be optimized instead of actual processor architectures. However, some results indicate that even larger speed-ups are likely to be achieved when trying to estimate how the compiler allocates registers to eliminated subexpressions. Figure 6.19 (see page 156) shows a large increase of data cache accesses and misses after advanced code hoisting. The reason for this behavior is that the newly generated local variables storing the common subexpressions can not be kept in the processor's registers. The compiler needs to insert spill code so that the eliminated expressions are frequently fetched from background memory.

In order to estimate at a high-level whether it is beneficial to eliminate a given expression, the following criteria can be analyzed:

Complexity: The expression $(3-y\%3-(y-2)\%3)*3$ is much more complex than $y+1$ so that the elimination tends to be more beneficial. In this example, this expression is so complex that an elimination is advantageous even if it is spilled to memory, since a memory access is usually cheaper than two modulo computations.

Reuse factor: If an expression occurs very frequently, it is a better candidate for elimination than an expression occurring only twice, because the local variable used to store the eliminated expression is reused more often.

Distance: If occurrences of a common subexpression are widely spread over a large range of code, it might not be a good idea to eliminate these occurrences and store them in a single variable. This is due to the fact that this new local variable would have a very long lifetime causing trouble to the register allocator. It might be better to group those occurrences of an expression which are very close to each other, and to eliminate only such a group using a new variable. This way, lifetimes of local variables covering long distances in the code can be avoided.

Register pressure: For every occurrence of a common subexpression in the code, a high-level estimation of the current register pressure should be performed in order to decide whether the elimination is useful or not.

The study of the effects of these individual factors on a set of actual processors and compilers is part of ongoing work. In the future, a high-level estimation mechanism will be developed combining the above factors and steering the elimination of common subexpressions. During this process, existing trade-offs imposed by the different criteria (e. g. when is the elimination of a complex but widely scattered expression beneficial) need to be explored.

Ring Buffer Replacement

Concerning ring buffer replacement, the full automation of the techniques proposed in chapter 7 is the most important part of the future work.

After that, a detailed study of code size and speed trade-offs needs to be performed. In the case of ring buffer replacement, this trade-off can be reduced to the question whether to apply loop unrolling for ring buffers or not. In order to take this decision, the effect of loop unrolling for ring buffers on different processor architectures needs to be studied first. The experimental results provided in chapter 7.3 indicate that the influence of this step on the measured runtimes depends on the actual processor.

Appendix A
Experimental Comparison of SUIF and IR-C/LANCE

In this appendix, all data measured during the experimental comparison of the high-SUIF and IR-C intermediate representations (see section 4.4) is listed. The following tables contain the runtimes of the benchmarks on the considered processors. For the Sun, Pentium Pro and TriMedia processors, these values represent CPU seconds, whereas the number of clock cycles is given for the TI C6x. Table A.1 shows the runtimes of the original benchmark versions. The execution times of these programs after being translated to high-SUIF and back to C code are listed in table A.2. Finally, table A.3 contains the corresponding data based on IR-C and the generation of data flow trees.

Original	CAVITY	ME	QSDPCM
Sun [sec]	0.66	2.49	1.94
Pentium Pro [sec]	0.36	1.21	3.39
TriMedia [sec]	4.28	18.52	10.16
TI C6x [cyc]	1,762,368,530	11,296,858,034	11,924,371,459

Table A.1.   Runtimes of original Benchmark Versions

SUIF	CAVITY	ME	QSDPCM
Sun [sec]	0.67	2.50	1.91
Pentium Pro [sec]	0.39	1.21	3.39
TriMedia [sec]	4.28	18.52	12.42
TI C6x [cyc]	1,796,919,290	11,296,857,420	12,391,188,263

Table A.2.   Runtimes of SUIF Benchmark Versions

IR-C	CAVITY	ME	QSDPCM
Sun *[sec]*	0.95	3.87	2.16
Pentium Pro *[sec]*	0.73	2.01	3.52
TriMedia *[sec]*	7.40	29.98	13.97
TI C6x *[cyc]*	2,221,046,857	11,776,884,602	12,718,918,691

Table A.3.   Runtimes of IR-C Benchmark Versions

Appendix B
Benchmarking Data for Loop Nest Splitting

In this appendix, all data measured during the benchmarking of loop nest splitting is listed. All results given in chapter 5 can be reproduced from the material presented in the following sections. In section B.1, the data for rating the behavior of the instruction pipelines and caches is given. Section B.2 lists the exact runtimes and code sizes of the benchmarks for all processors considered in section 5.5.1.2. All data related to the energy dissipation of the ARM7TDMI embedded RISC core is presented in section B.3. Finally, the measured results relating to the combined application of data partitioning and loop nest splitting are given in section B.4.

B.1   Values of performance-monitoring Counters
B.1.1   Intel Pentium III

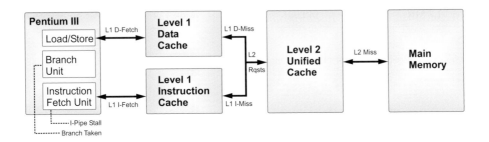

Figure B.1.   Structure of the Intel Pentium III Architecture

Following the definition given in [Hennessy and Patterson, 2003], the Intel Pentium III processor is a typical Harvard architecture having a single main memory but separate instruction and data caches as illustrated in figure B.1. Internally, the processor communicates with its level 1 caches via separate data and instruction buses. The on-chip level 1 data and instruction caches each have a size of 16 kB and use a 4-way set-associative mapping with 32 bytes per cache line.

191

The off-chip level 2 cache is a unified cache storing both data and instructions. It has a size of 256 kB with 32 bytes per line using an 8-way set-associative mapping.

This section contains all pipeline and cache related counter values determined with the help of the PMC library [Heller, 2000] for the Intel Pentium III processor. For obtaining these results, the benchmarks were compiled and linked with this library which makes use of hardware performance-monitoring counters of the Pentium processor [Intel Corp., 2002]. This way, reliable values can be generated by profiling without using erroneous cache simulation software. The probe-points where the data listed in the following tables was measured within the Pentium III architecture are indicated in figure B.1.

In order to avoid side-effects of other running processes being measured during the execution of the benchmarks, all periodically active processes were terminated beforehand. Furthermore, each measurement was performed twenty times. After that, the minimum and maximum values measured were discarded and average values over the remaining eighteen measurements were computed. This approach leads to very stable results throughout all performed measurements. For example, the measured numbers of taken branches for the CAVITY benchmark have a deviation of less than 0.2%.

Table B.1 shows the average values for the original versions of the benchmarks, whereas the results after loop nest splitting are given in table B.2. Out of these values, the diagram shown in figure 5.12 (see page 103) was generated.

Original	CAVITY	ME	QSDPCM
Cycles	457,882,361	1,057,369,204	416,172,096
Branch Taken	15,503,809	209,133,207	35,145,643
I-Pipe Stall	3,724,967	4,269,815	5,256,488
L1 I-Fetch	453,220,737	1,051,974,908	406,400,954
L1 I-Miss	10,122	17,424	39,100
L1 D-Fetch	277,196,342	597,853,395	144,444,058
L1 D-Miss	136,056	43,693	46,332
L2 Requests	113,894	86,109	157,424
L2 Miss	148,941	69,129	58,746

Table B.1.   PMC Values for original Benchmark Versions

Loop Nest Split	CAVITY	ME	QSDPCM
Cycles	312,217,838	279,831,507	304,873,722
Branch Taken	14,248,396	56,755,097	26,928,781
I-Pipe Stall	3,338,489	1,858,695	3,994,489
L1 I-Fetch	308,599,829	277,828,106	297,909,549
L1 I-Miss	8,626	5,494	28,699
L1 D-Fetch	232,823,980	87,489,875	150,062,180
L1 D-Miss	126,279	31,809	38,419
L2 Requests	99,009	39,770	121,331
L2 Miss	134,209	38,857	48,418

Table B.2.   PMC Values for Benchmarks after Loop Nest Splitting

B.1.2   Sun UltraSPARC III

Barring the presence of a prefetch cache, the architecture of the Sun UltraSPARC III processor [Sun Microsystems Inc., 2002] is principally comparable to that of the Pentium III since it also consists of separate level 1 data and instruction caches and a unified level 2 cache (cf. figure B.2). As in the case of the Pentium, this CPU allows processor event counting for performance measurements.

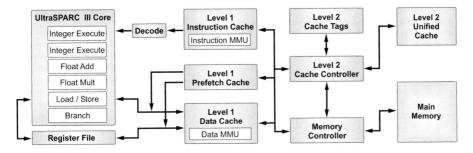

Figure B.2.   Structure of the Sun UltraSPARC III Architecture

The main difference between these processors is the size of the various memories. The on-chip level 1 instruction cache of the UltraSPARC III has a size of 32 kB, the level 1 data cache is 64 kB large. Both caches use a 4-way set associative mapping with 32 bytes per cache line. The external level 2 cache has a size of 8 MB with 512 bytes per line using a 2-way set-associative mapping. The Sun processor has a very large register file containing 160 general purpose integer registers and up to 32 floating-point registers. The floating-point registers can be loaded automatically by using the prefetch cache and its associated speculative hardware controller[1].

The pipeline and cache related material presented in section 5.5.1.1 for the Sun UltraSPARC III processor was collected with the help of the performance instrumentation counters (*PIC*) [Sun Microsystems Inc., 2002] of the processor and the cputrack utility of the Solaris operating system. Each measurement was performed twenty times. Afterwards, the minimum and maximum values measured were discarded and average values over the remaining eighteen measurements were computed.

Table B.3 shows the average values for the original versions of the benchmarks, whereas the results after loop nest splitting are given in table B.4. The diagram shown in figure 5.13 (see page 105) was generated from these values.

Original	*CAVITY*	*ME*	*QSDPCM*
Cycles	393,472,375	868,531,375	237,383,891
Branch Taken	26,211,162	147,832,045	27,915,617
I-Pipe Stall	13,157,318	34,530,345	6,123,507

---

[1]Since floating-point computations are not performed by the applications serving as benchmarks in this work, the UltraSPARC III prefetch cache is not taken into account during performance monitoring.

Original	CAVITY	ME	QSDPCM
L1 I-Fetch	147,889,880	412,710,424	128,183,341
L1 I-Miss	18,293	15,786	11,767
L1 D-Fetch	89,690,758	6,007,608	24,754,265
L1 D-Miss	259,081	168,838	140,497
L2 Requests	66,330,936	5,812,883	24,571,642
L2 Miss	25,121	10,179	3,558

Table B.3. PIC Values for original Benchmark Versions

Loop Nest Split	CAVITY	ME	QSDPCM
Cycles	262,503,992	210,301,284	133,130,265
Branch Taken	27,598,854	17,314,895	12,836,237
I-Pipe Stall	9,335,217	9,300,382	2,727,359
L1 I-Fetch	106,189,973	71,526,680	67,382,877
L1 I-Miss	18,778	14,139	11,760
L1 D-Fetch	70,652,809	5,903,200	25,050,442
L1 D-Miss	126,011	163,925	128,130
L2 Requests	54,952,441	5,764,959	24,655,932
L2 Miss	24,850	7,265	3,476

Table B.4. PIC Values for Benchmarks after Loop Nest Splitting

B.1.3   MIPS R10000

With respect to the first and second level caches, the architecture of the MIPS R10000 processor is equivalent to the one of the Intel Pentium III. In the case of the MIPS, the sizes of the level 1 instruction and data caches are 32 kB. Both caches are 2-way set-associative, the block size of the instruction cache is 16 words, the data cache uses 8 words per block. The secondary unified cache has a size of 1 MB and is also 2-way set-associative.

For accessing the MIPS R10000 performance counters (*PC*) [MIPS Technologies Inc., 1996], the perfex utility was used. In analogy to the previous sections, every measurement was performed twenty times in order to obtain stable results. The following tables B.5 and B.6 show the average results of these measurements after removal of the minimum and maximum values.

Original	CAVITY	ME	QSDPCM
Cycles	364,843,119	729,839,198	124,772,234
Branch Instrs	55,231,492	249,296,813	53,268,372
L1 I-Fetch	576,065,664	1,115,530,132	282,093,449
L1 I-Miss	61,616	51,843	2,947
L1 D-Fetch	199,546,246	39,341,594	24,406,526

Original	CAVITY	ME	QSDPCM
L1 D-Miss	155,691	41,550	7,396
L2 Requests	217,277	93,376	10,347
L2 Miss	13,922	11,879	436

Table B.5.   PC Values for original Benchmark Versions

Loop Nest Split	CAVITY	ME	QSDPCM
Cycles	269,520,370	285,561,094	45,706,635
Branch Instrs	4,532,214	40,149,316	17,973,082
L1 I-Fetch	394,699,884	211,966,736	97,632,479
L1 I-Miss	52,555	22,365	1,887
L1 D-Fetch	126,904,539	21,892,341	7,985,215
L1 D-Miss	131,260	35,238	4,819
L2 Requests	183,898	57,647	6,707
L2 Miss	10,442	1,558	451

Table B.6.   PC Values for Benchmarks after Loop Nest Splitting

B.2   Execution Times and Code Sizes

In this section, the absolute runtimes and code sizes of the benchmarks are given. As mentioned in section 3.3.2, these values are obtained by compilation and execution of the benchmarks on a large set of different processors. Compilation is always performed using the highest degree of optimization provided by the corresponding compiler. The compilers used for the different processors are listed in the following table B.7 together with their command line options.

Processor	Compiler	Version	Command Line Options
Sun UltraSPARC III	Sun WorkShop	6.2	CC -fast -xtarget=ultra3 -xarch=v9b
Intel Pentium III	GNU gcc	2.95.3	gcc -O3
HP-PA 9000/785	HP Softbench	A.01.18	aCC +O4
MIPS R10000	MIPSpro	7.2.1	CC -Ofast
Power-PC G4	GNU gcc	2.95.4	gcc -O3
DEC Alpha EV4	GNU gcc	2.95.4	gcc -O3
TriMedia-1000	Philips SDE	V5.7.1	tmcc -O5 -host WinNT
TI TMS320C62	TI CodeGen Tools	3.01	cl6x -o3
ARM7TDMI thumb	ARM SDT	2.50	tcc -O2 -Otime
ARM7TDMI arm	ARM SDT	2.50	armcc -O2 -Otime

Table B.7.   Compilers and Optimization Levels for Runtime and Code Size Measurements

The values given in the following tables denote the number of CPU seconds and the number of clock cycles resp. required for the execution of a benchmark. Table B.8 shows the runtimes before the optimization. The speed-ups reported in figure 5.15 (see page 107) were generated by comparing this data with the runtimes after loop nest splitting which can be found in table B.9.

Original	CAVITY	ME	QSDPCM
Sun *[cyc]*	393,472,375	868,531,375	237,383,891
Pentium *[cyc]*	457,882,361	1,057,369,204	416,172,096
HP *[sec]*	0.56	1.87	0.30
MIPS *[cyc]*	364,843,119	729,839,198	124,772,234
Power-PC G4 *[sec]*	0.82	2.41	0.48
DEC Alpha *[sec]*	4.61	12.76	3.46
TriMedia *[sec]*	5.10	21.92	3.38
TI C6x *[cyc]*	1,264,823,138	2,681,116,485	1,237,341,543
ARM7 thumb *[cyc]*	146,308,938	8,230,155,008	2,137,855,898
ARM7 arm *[cyc]*	172,938,892	9,663,200,190	2,005,827,480

Table B.8.   Runtimes of original Benchmark Versions

Loop Nest Split	CAVITY	ME	QSDPCM
Sun *[cyc]*	262,503,992	210,301,284	133,130,265
Pentium *[cyc]*	312,217,838	279,831,507	304,873,722
HP *[sec]*	0.36	0.55	0.12
MIPS *[cyc]*	269,520,370	285,561,094	45,706,635
Power-PC G4 *[sec]*	0.58	0.61	0.46
DEC Alpha *[sec]*	2.99	6.01	2.60
TriMedia *[sec]*	3.89	13.93	2.74
TI C6x *[cyc]*	1,167,015,404	733,300,141	867,214,454
ARM7 thumb *[cyc]*	122,616,256	3,305,386,900	1,891,857,950
ARM7 arm *[cyc]*	152,914,152	4,367,474,972	1,840,264,050

Table B.9.   Runtimes of Benchmarks after Loop Nest Splitting

In the following tables B.10 and B.11, the code sizes before and after loop nest splitting for the processors listed above are given. The values listed in these tables denote the number of assembly instructions emitted by the used compilers.

Original	CAVITY	ME	QSDPCM
Sun	634	369	165
Pentium	521	249	145
HP	444	492	176

Original	CAVITY	ME	QSDPCM
MIPS	691	456	242
Power-PC G4	594	264	160
DEC Alpha	615	302	143
TriMedia	1,750	960	585
TI C6x	577	295	184
ARM7 thumb	601	300	273
ARM7 arm	469	216	136

Table B.10.   Code Sizes of original Benchmark Versions

Loop Nest Split	CAVITY	ME	QSDPCM
Sun	951	418	303
Pentium	898	335	273
HP	673	745	224
MIPS	931	498	263
Power-PC G4	964	325	269
DEC Alpha	1,124	387	254
TriMedia	2,580	1,235	825
TI C6x	1,042	373	371
ARM7 thumb	1,010	437	485
ARM7 arm	784	282	215

Table B.11.   Code Sizes of Benchmarks after Loop Nest Splitting

B.3   Energy Consumption of an ARM7TDMI Core

This section contains the material gathered when using an accurate instruction-level energy model [Steinke et al., 2001a] for an ARM7 embedded RISC core during benchmarking. The first four rows of the following tables denote the absolute number of different kinds of memory accesses. In the lower part of these tables, the energy dissipation of the benchmarks is given in μJ. Table B.12 shows the data of the energy profiling before loop nest splitting, whereas table B.13 gives the values after the source code optimization.

Original	CAVITY	ME	QSDPCM
Instruction Read	1,488,952,642	4,187,759,099	1,436,042,507
Data Read	171,633,720	800,612,554	397,077,933
Data Write	36,299,715	95,364,299	29,922,359
Total Memory Accesses	1,696,886,077	5,083,735,952	1,863,042,799

Original	CAVITY	ME	QSDPCM
Memory Energy Consumption [μJ]	44,051,984.515	143,911,901.100	54,457,644.234
CPU Core Energy Consumption [μJ]	19,996,772.345	63,456,704.577	24,112,184.733
Total Energy Consumption [μJ]	64,048,756.860	207,368,605.677	78,569,828.967

Table B.12.   Energy Consumption of original Benchmark Versions

Loop Nest Split	CAVITY	ME	QSDPCM
Instruction Read	1,138,610,481	1,804,375,600	874,713,006
Data Read	159,407,055	278,139,796	205,511,633
Data Write	45,190,087	93,331,459	1,369,659
Total Memory Accesses	1,343,207,623	2,175,846,855	1,081,594,298
Memory Energy Consumption [μJ]	35,406,147.932	60,858,791.538	30,364,170.438
CPU Core Energy Consumption [μJ]	16,320,808.977	27,015,207.376	13,489,391.884
Total Energy Consumption [μJ]	51,726,956.909	87,873,998.913	43,853,562.322

Table B.13.   Energy Consumption of Benchmarks after Loop Nest Splitting

B.4   Combined Data Partitioning and Loop Nest Splitting
B.4.1   Execution Times and Code Sizes

Using the energy profiling methodology described in section 3.3.3, the behavior of benchmarks with respect to execution times and code sizes was measured. In the following tables B.14 and B.15, this data which is used for the generation of figures 5.18 and 5.19 is listed. The rows of these tables denote the results obtained for the original benchmarks and for the code versions after data partitioning and after combined data partitioning and loop nest splitting respectively. Runtimes are given in terms of the number of CPU cycles, code sizes are represented by the number of generated assembly instructions.

	FIR	INS	ME	SELS
Original	555,807	768,265	289,566,525	1,174,067
Data Partitioning	646,941	1,050,555	207,809,649	1,275,674
Loop Nest Splitting	542,173	842,793	143,190,815	1,156,865

Table B.14.   Runtimes of Benchmarks after Data Partitioning and Loop Nest Splitting

	FIR	INS	ME	SELS
Original	92	92	644	80
Data Partitioning	148	204	544	144
Loop Nest Splitting	108	296	1,072	212

Table B.15.   Code Sizes of Benchmarks after Data Partitioning and Loop Nest Splitting

B.4.2   Energy Consumption

In the following tables B.16, B.17 and B.18, the energy consumption measured for the considered benchmarks is presented in μJ. These tables list the individual energy consumptions measured for the memories, the ARM7 processor and the total combination of both.

Original	FIR	INS	ME	SELS
Memory $[\mu J]$	1,666.966	2,676.981	1,869,645.112	3,266.707
Processor $[\mu J]$	2,626.354	3,591.574	1,322,463.897	5,350.872
Total $[\mu J]$	4,293.319	6,268.555	3,192,109.009	8,577.579

Table B.16.   Energy Consumption of original Benchmark Versions

Data Partitioning	FIR	INS	ME	SELS
Memory $[\mu J]$	1,055.054	523.079	1,139,150.980	517.677
Processor $[\mu J]$	2,995.395	4,695.862	943,968.163	5,671.097
Total $[\mu J]$	4,050.450	5,218.941	2,083,119.140	6,188.774

Table B.17.   Energy Consumption of Benchmarks after Data Partitioning

Loop Nest Split	FIR	INS	ME	SELS
Memory $[\mu J]$	889.407	484.249	1,162,093.962	501.542
Processor $[\mu J]$	2,533.578	3,798.669	649,427.705	5,163.372
Total $[\mu J]$	3,422.985	4,282.919	1,811,521.667	5,664.914

Table B.18.   Energy Consumption of Benchmarks after Loop Nest Splitting

Appendix C
Benchmarking Data for Advanced Code Hoisting

This appendix contains all benchmarking data related to the advanced code hoisting source code optimization that is presented in chapter 6. Section C.1 contains all material concerning instruction pipeline and cache behavior. In section C.2, tables representing the execution times and code sizes before and after advanced code hoisting are given. Finally, the numerical results of the energy measurements are presented in section C.3.

C.1   Values of performance-monitoring Counters
C.1.1   Intel Pentium III

This section contains all values gathered during the analysis of the CAVITY and QSDPCM benchmarks using the PMC counters [Heller, 2000] of an Intel Pentium III processor. Table C.1 lists all measured pipeline and cache events before advanced code hoisting, whereas the optimized benchmarks lead to the numbers given in table C.2. This data was used to generate figure 6.18 on page 155.

Original	CAVITY	QSDPCM
Cycles	886,044,280	1,320,066,815
I-Pipe Stall	8,288,841	5,475,499
L1 I-Fetch	876,066,628	1,312,754,126
L1 I-Miss	46,067	62,684
L1 D-Fetch	454,632,809	198,137,124
L1 D-Miss	244,753	76,850
L2 Requests	305,044	283,194
L2 Miss	309,652	139,908

Table C.1.   PMC Values for original Benchmark Versions

Advanced Code Hoisting	CAVITY	QSDPCM
Cycles	451,200,023	887,891,544
I-Pipe Stall	4,763,297	4,125,372
L1 I-Fetch	445,661,391	881,717,548
L1 I-Miss	23,455	44,832
L1 D-Fetch	279,636,663	147,404,416
L1 D-Miss	185,075	59,789
L2 Requests	195,902	186,774
L2 Miss	212,446	106,318

Table C.2. PMC Values for Benchmarks after Advanced Code Hoisting

C.1.2   Sun UltraSPARC III

The following tables C.3 and C.4 provide the detailed results of the measurement of cache and pipeline behavior for the Sun UltraSPARC III processor. Again, these tables represent the benchmarks' behavior before and after advanced code hoisting. The graphical depiction of these tables is given in figure 6.19 (see page 156).

Original	CAVITY	QSDPCM
Cycles	919,788,987	441,535,260
I-Pipe Stall	4,379,471	1,044,421
L1 I-Fetch	211,577,408	127,452,667
L1 I-Miss	26,179	16,853
L1 D-Fetch	83,301,198	64,401,285
L1 D-Miss	161,422	225,041
L2 Requests	61,808,058	10,508,983
L2 Miss	27,296	6,635

Table C.3. PIC Values for original Benchmark Versions

Advanced Code Hoisting	CAVITY	QSDPCM
Cycles	585,166,843	313,587,942
I-Pipe Stall	6,815,793	933,159
L1 I-Fetch	159,798,164	98,498,105
L1 I-Miss	23,300	17,145
L1 D-Fetch	138,752,973	46,119,925
L1 D-Miss	1,847,646	271,568
L2 Requests	77,474,196	10,518,676
L2 Miss	26,749	6,320

Table C.4. PIC Values for Benchmarks after Advanced Code Hoisting

C.1.3   MIPS R10000

The outcome of the use of the MIPS R10000 performance counters [MIPS Technologies Inc., 1996] in the context of advanced code hoisting is documented in the following tables. Table C.5 denotes the 100% base line of figure 6.20 (cf. page 157), whereas the behavior of the benchmarks after optimization is shown in table C.6.

Original	CAVITY	QSDPCM
Cycles	1,125,723,677	640,338,861
L1 I-Fetch	709,125,257	796,989,881
L1 I-Miss	95,004	48,072
L1 D-Fetch	150,698,118	122,368,706
L1 D-Miss	124,772	37,174
L2 Requests	219,753	85,494
L2 Miss	1,813	2,873

Table C.5.   PC Values for original Benchmark Versions

Advanced Code Hoisting	CAVITY	QSDPCM
Cycles	749,360,574	437,667,273
L1 I-Fetch	608,237,085	546,016,225
L1 I-Miss	60,015	38,809
L1 D-Fetch	130,268,531	63,797,329
L1 D-Miss	107,195	24,350
L2 Requests	167,207	63,187
L2 Miss	2,134	1,476

Table C.6.   PC Values for Benchmarks after Advanced Code Hoisting

C.2   Execution Times and Code Sizes

The values given in the following tables denote the number of CPU seconds and the number of clock cycles resp. required for the execution of a benchmark. Table C.7 shows the runtimes before the optimization. The speed-ups reported in figure 6.21 (see page 158) were generated by comparing this data with the runtimes after advanced code hoisting which can be found in table C.8.

Original	CAVITY	QSDPCM
Sun *[cyc]*	919,788,987	441,535,260
Pentium *[cyc]*	886,044,280	1,320,066,815
HP *[sec]*	1.90	0.66

Original	CAVITY	QSDPCM
MIPS *[cyc]*	1,125,723,677	640,338,861
TriMedia *[sec]*	16.35	8.14
TI C6x *[cyc]*	9,153,766,452	5,686,182,014
ARM7 thumb *[cyc]*	615,668,392	5,364,778,088
ARM7 arm *[cyc]*	545,809,832	5,465,169,956

Table C.7.   Runtimes of original Benchmark Versions

Advanced Code Hoisting	CAVITY	QSDPCM
Sun *[cyc]*	585,166,843	313,587,942
Pentium *[cyc]*	451,200,023	887,891,544
HP *[sec]*	0.95	0.61
MIPS *[cyc]*	749,360,574	437,667,273
TriMedia *[sec]*	7.27	5.13
TI C6x *[cyc]*	4,088,968,070	5,553,741,219
ARM7 thumb *[cyc]*	187,031,290	3,671,444,668
ARM7 arm *[cyc]*	219,956,270	3,681,660,444

Table C.8.   Runtimes of Benchmarks after Advanced Code Hoisting

The values given in the following two tables C.9 and C.10 denote the number of assembly instructions emitted by compilers for all the processors considered in this section. These values represent the code sizes of the benchmark applications and were used to generate figure 6.23.

Original	CAVITY	QSDPCM
Sun	759	3,640
Pentium	781	3,184
HP	619	3,337
MIPS	618	5,238
TriMedia	1,505	11,700
TI C6x	661	2,977
ARM7 thumb	672	2,815
ARM7 arm	502	1,951

Table C.9.   Code Sizes of original Benchmark Versions

Advanced Code Hoisting	*CAVITY*	*QSDPCM*
Sun	726	3,678
Pentium	537	2,977
HP	470	2,237
MIPS	566	5,035
TriMedia	1,220	8,430
TI C6x	513	2,911
ARM7 thumb	493	2,631
ARM7 arm	399	1,860

Table C.10.   Code Sizes of Benchmarks after Advanced Code Hoisting

C.3   Energy Consumption of an ARM7TDMI Core

The following tables C.11 and C.12 show the energy consumption of the CAVITY and QSDPCM applications before and after advanced code hoisting. The energy consumption is given in μJ. In addition, the numbers of read and write accesses to memory are shown in order to isolate the influence of instruction and data memory accesses on the total energy dissipation reduction. Using the material given in this appendix, figure 6.24 was generated.

Original	*CAVITY*	*QSDPCM*
Instruction Read	6,588,940,240	3,784,874,336
Data Read	278,214,814	262,449,252
Data Write	156,745,213	173,823,334
Total Memory Accesses	7,023,900,267	4,221,146,922
Memory Energy Consumption $[\mu J]$	176,658,574.605	189,172,638.607
CPU Core Energy Consumption $[\mu J]$	63,178,793.311	60,739,680.640
Total Energy Consumption $[\mu J]$	239,837,367.916	249,912,319.247

Table C.11.   Energy Consumption of original Benchmark Versions

Advanced Code Hoisting	*CAVITY*	*QSDPCM*
Instruction Read	1,418,068,251	2,532,164,312
Data Read	180,337,570	174,988,459
Data Write	62,785,754	127,906,337
Total Memory Accesses	1,661,191,575	2,835,059,108
Memory Energy Consumption $[\mu J]$	43,931,945.839	126,060,155.330
CPU Core Energy Consumption $[\mu J]$	17,170,978.995	40,338,055.840
Total Energy Consumption $[\mu J]$	61,102,924.834	166,398,211.170

Table C.12.   Energy Consumption of Benchmarks after Advanced Code Hoisting

Appendix D
Benchmarking Data for Ring Buffer Replacement

In this appendix, the numerical data collected during the benchmarking of the CAVITY application before and after ring buffer replacement (cf. chapter 7) is presented. Like in the previous appendices, values of performance measuring counters are given first in section D.1. The execution times and code sizes of CAVITY are shown in section D.2, and section D.3 contains the material related to the energy consumption of the benchmark application.

D.1   Values of performance-monitoring Counters
D.1.1   Intel Pentium III

Table D.1 shows the behavior of the Intel Pentium's performance counters before and after ring buffer replacement applied to the CAVITY benchmark. Since this source code optimization consists of two individual steps, the table contains the corresponding material for both steps. The contents of table D.1 lead to figure 7.10.

CAVITY	Original	Scalarization	Loop Unrolling
Cycles	376,926,297	301,924,317	298,032,656
Branch Taken	32,202,660	32,170,990	31,327,466
L1 I-Fetch	370,839,349	297,357,526	293,208,232
L1 I-Miss	21,468	17,090	18,626
L1 D-Fetch	229,747,974	125,953,229	140,614,410
L1 D-Miss	202,183	156,841	160,443
L2 Requests	189,164	149,597	156,032
L2 Miss	251,701	185,395	194,622

Table D.1.   PMC Values of Ring Buffer Replacement for CAVITY

D.1.2   Sun UltraSPARC III

In analogy to the previous section, table D.2 lists the values measured during the benchmarking of the pipeline and cache behavior of the Sun processor. Figure 7.11 is based on this data.

CAVITY	Original	Scalarization	Loop Unrolling
Cycles	432,349,319	352,897,846	339,285,738
Branch Taken	14,986,737	15,011,327	14,107,032
L1 I-Fetch	111,471,729	89,330,442	82,105,664
L1 I-Miss	23,838	20,377	26,411
L1 D-Fetch	89,839,813	42,497,246	40,970,601
L1 D-Miss	151,192	144,067	142,002
L2 Requests	55,659,331	29,958,222	23,760,856
L2 Miss	25,129	25,189	25,236

Table D.2.   PIC Values of Ring Buffer Replacement for CAVITY

D.1.3   MIPS R10000

The generation of figure 7.12 was done using the following values for the MIPS R10000 processor.

CAVITY	Original	Scalarization	Loop Unrolling
Cycles	388,558,307	267,475,612	264,119,956
Branch Instrs	1,810,186	1,810,710	899,853
L1 I-Fetch	303,596,581	222,716,827	213,782,930
L1 I-Miss	6,722	5,821	6,644
L1 D-Fetch	54,240,917	23,640,372	38,006,950
L1 D-Miss	45,393	40,040	41,624
L2 Requests	52,120	45,868	48,281
L2 Miss	1,240	1,102	1,088

Table D.3.   PC Values of Ring Buffer Replacement for CAVITY

D.2   Execution Times and Code Sizes

Table D.4 contains the runtimes of the CAVITY application before and after ring buffer replacement for the processors considered in figure 7.13.

CAVITY	Original	Scalarization	Loop Unrolling
Sun *[cyc]*	432,349,319	352,897,846	339,285,738
Pentium *[cyc]*	376,926,297	301,924,317	298,032,656

CAVITY	Original	Scalarization	Loop Unrolling
HP [sec]	0.57	0.58	0.49
MIPS [cyc]	388,558,307	267,475,612	264,119,956
Power-PC G4 [sec]	0.67	0.47	0.46
TriMedia [sec]	12.09	10.92	11.41
TI C6x [cyc]	4,128,585,179	3,676,253,665	3,568,152,349
ARM7 thumb [cyc]	166,032,066	126,265,882	112,714,688
ARM7 arm [cyc]	185,206,096	152,739,656	140,734,138

Table D.4.   Runtimes of Ring Buffer Replacement for CAVITY

In table D.5, the total amount of assembly statements emitted by compilers for the processors studied in section 7.3.2 is shown. These values represent the code sizes of the CAVITY benchmark and were used to create figure 7.14.

CAVITY	Original	Scalarization	Loop Unrolling
Sun	1,142	864	1,335
Pentium	893	713	1,191
HP	760	653	1,080
MIPS	914	726	1,070
Power-PC G4	777	617	935
TriMedia	1,976	1,796	2,705
TI C6x	871	767	1,164
ARM7 thumb	822	670	1,054
ARM7 arm	675	557	889

Table D.5.   Code Sizes of Ring Buffer Replacement for CAVITY

D.3   Energy Consumption of an ARM7TDMI Core

The material generated during the energy measurements for ring buffer replacement is presented in table D.6. The graphical representation of this data is given in figure 7.15.

CAVITY	Original	Scalarization	Loop Unrolling
Instruction Read	1,978,556,017	1,580,927,829	1,546,459,733
Data Read	143,336,875	107,605,135	90,668,705
Data Write	67,523,019	64,965,929	50,584,058
Total Memory Accesses	2,189,415,911	1,753,498,893	1,687,712,496

CAVITY	Original	Scalarization	Loop Unrolling
Memory Energy Consumption $[\mu J]$	55,733,106.652	45,550,263.420	43,296,308.157
CPU Core Energy Consumption $[\mu J]$	20,528,332.040	16,769,564.973	15,660,639.037
Total Energy Consumption $[\mu J]$	76,261,438.692	62,319,828.393	58,956,947.194

Table D.6.   Energy Consumption of Ring Buffer Replacement for CAVITY

References

[3GPP, 1999] 3GPP (1999). *ANSI-C Code for the adaptive Multirate Speech Codec.* 3rd Generation Partnership Project, Literature Number 3G TS 26.073 V2.0.0.

[Aho et al., 1989] Aho, A. V., Ganapathi, M., and Tjiang, S. W. K. (1989). Code Generation using Tree Matching and Dynamic Programming. *ACM Transactions on Programming Languages and Systems*, 11(4).

[Aho et al., 1986] Aho, A. V., Sethi, R., and Ullman, J. D. (1986). *Compilers. Principles, Techniques, and Tools.* Addison-Wesley, Reading.

[ARM Ltd., 2001] ARM Ltd. (2001). *ARM7TDMI Technical Reference Manual.* Advanced RISC Machines Ltd., Document Number ARM DDI 0029G.

[Bäck, 1996] Bäck, T. (1996). *Evolutionary Algorithms in Theory and Practice.* Oxford University Press.

[Bacon et al., 1994] Bacon, D. F., Graham, S. L., and Sharp, O. J. (1994). Compiler Transformations for High-Performance Computing. *ACM Computing Surveys*, 26(4):345–420.

[Banakar et al., 2002] Banakar, R., Steinke, S., Lee, B.-S., Balakrishnan, M., and Marwedel, P. (2002). Scratchpad Memory: A Design Alternative for Cache on-chip Memory in Embedded Systems. In *Proceedings of "International Symposium on Hardware/Software Codesign" (CODES)*, pages 73–78, Estes Park.

[Bashford and Leupers, 1999] Bashford, S. and Leupers, R. (1999). Phase-Coupled Mapping of Data Flow Graphs to irregular Data Paths. *Design Automation for Embedded Systems*, 4(2/3):1–50.

[Bastoul, 2003] Bastoul, C. (2003). *Generating Loops for scanning Polyhedra: CLooG User's Guide.* Paris.

[Bister et al., 1989] Bister, M., Taeymans, Y., and Cornelis, J. (1989). Automatic Segmentation of cardiac MR Images. *IEEE Journal on Computers in Cardiology*, pages 215–218.

[Boekhold et al., 1999] Boekhold, M., Karkowski, I., and Corporaal, H. (1999). Transforming and parallelizing ANSI C Programs using Pattern Recognition. *High Performance Computing and Networking Conference, Amsterdam*, pages 673–682.

211

[Brockmeyer et al., 2003] Brockmeyer, E., Miranda, M., Corporaal, H., and Catthoor, F. (2003). Layer Assignment Techniques for low Energy in multi-layered Memory Organisations. In *Proceedings of "Design, Automation and Test in Europe" (DATE)*, pages 1070–1075, Munich.

[Catthoor et al., 2002] Catthoor, F., Danckaert, K., Kulkarni, C., Brockmeyer, E., Kjeldsberg, P. G., van Achteren, T., and Omnes, T. (2002). *Data Access and Storage Management for embedded programmable Processors*. Kluwer Academic Publishers, Boston.

[Catthoor et al., 1998] Catthoor, F., Wuytack, S., De Greef, E., et al. (1998). *Custom Memory Management Methodology – Exploration of Memory Organisation for embedded Multimedia System Design*. Kluwer Academic Publishers.

[Clauss and Loechner, 1998] Clauss, P. and Loechner, V. (1998). Parametric Analysis of polyhedral Iteration Spaces. *Journal of VLSI Signal Processing*, 19(2):179–194.

[Cooper et al., 1999] Cooper, K. D., Schielke, P. J., and Subramanian, D. (1999). Optimizing for reduced Code Space using Genetic Algorithms. In *Proceedings of "Workshop on Languages, Compilers, and Tools for Embedded Systems" (LCTES)*, pages 1–9, Atlanta.

[Coors et al., 1999] Coors, M., Wahlen, O., Keding, H., Lüthje, O., and Meyr, H. (1999). TI C62x Performance Code Optimization (in German). In H. Rogge, R. E., editor, *DSP Deutschland '99 - Grundlagen, Architekturen, Tools, Applikationen*, pages 155–164, Poing. Design & Elektronik.

[Danckaert et al., 1999] Danckaert, K., Catthoor, F., and De Man, H. (1999). Platform independent Data Transfer and Storage Exploration illustrated on a parallel Cavity Detection Algorithm. In *Proceedings of "Parallel and Distributed Processing Techniques and Applications" (PDPTA)*, pages 1669–1675, Las Vegas.

[Daylight et al., 2002] Daylight, E. G., Wuytack, S., Ykman-Couvreur, C., and Catthoor, F. (2002). Analyzing Energy friendly Steady State Phases of dynamic Application Execution in Terms of sparse Data Structures. In *Proceedings of "International Symposium on Low Power Electronics and Design" (ISLPED)*, pages 76–79, Monterey.

[Drechsler, 1998] Drechsler, R. (1998). *Evolutionary Algorithms for VLSI CAD*. Kluwer Academic Publishers, Boston.

[Falk, 2002] Falk, H. (2002). Control Flow Optimization by Loop Nest Splitting at the Source Code Level. Research Report 773, University of Dortmund.

[Falk et al., 2003a] Falk, H., Ghez, C., Miranda, M., and Leupers, R. (2003a). High-level Control Flow Transformations for Performance Improvement of address-dominated Multimedia Applications. In *Proceedings of "Workshop on Synthesis and System Integration of Mixed Information Technologies" (SASIMI)*, pages 338–344, Hiroshima.

[Falk and Marwedel, 2003] Falk, H. and Marwedel, P. (2003). Control Flow driven Splitting of Loop Nests at the Source Code Level. In *Design, Automation and Test in Europe (DATE)*, pages 410–415, Munich.

[Falk et al., 2003b] Falk, H., Marwedel, P., and Catthoor, F. (2003b). *Control Flow driven Splitting of Loop Nests at the Source Code Level*, volume Embedded Software for SOC, chapter 17, pages 215–229. Kluwer Academic Publishers, Boston.

[Falk and Verma, 2004] Falk, H. and Verma, M. (2004). Combined Data Partitioning and Loop Nest Splitting for Energy Consumption Minimization. In *Proceedings of "International Workshop on Software and Compilers for Embedded Systems " (SCOPES)*, Amsterdam.

[Fraboulet et al., 1999] Fraboulet, A., Huard, G., and Mignotte, A. (1999). Loop Alignment for Memory Accesses Optimization. In *Proceedings of "International Symposium on System Synthesis" (ISSS)*, pages 71–77, San Jose.

[Garey and Johnson, 1979] Garey, M. R. and Johnson, D. S. (1979). *Computers and Intractability. A Guide to the Theory of NP-Completeness.* Freeman, New York, 1. edition.

[GCC, 2003] GCC (2003). The GCC Home Page, Free Software Foundation, Boston. *http://gcc.gnu.org.*

[Ghez et al., 2000] Ghez, C., Miranda, M., Vandecappelle, A., et al. (2000). Systematic high-level Address Code Transformations for piece-wise linear Indexing. In *Proceedings of "IEEE Workshop on Signal Processing Systems" (SIPS)*, pages 623–632, Lafayette.

[Givargis, 2004] Givargis, T. (2004). Embedded Systems Design Challenges. *http://www.ics.uci.edu/~lopes/teaching/ubicomp/givargis.ppt.*

[Greef et al., 1997] Greef, E. d., Catthoor, F., and Man, H. d. (1997). Memory Size Reduction through Storage Order Optimization for embedded parallel Multimedia Applications. *Parallel Computing*, 23(12):1811–1837.

[Gupta et al., 2000] Gupta, S., Miranda, M., Catthoor, F., et al. (2000). Analysis of high-level Address Code Transformations for programmable Processors. In *Proceedings of "Design, Automation and Test in Europe" (DATE)*, pages 9–13, Paris.

[Heller, 2000] Heller, D. (2000). Performance-monitoring Counters Library for Intel / AMD Processors and Linux. *http://www.scl.ameslab.gov/Projects/Rabbit/.*

[Hennessy and Patterson, 2003] Hennessy, J. L. and Patterson, D. A. (2003). *Computer Architecture: A quantitative Approach.* Morgan Kaufmann Publishers, San Francisco, 3. edition.

[Holland, 1992] Holland, J. H. (1992). *Adaption in natural and artificial Systems.* MIT Press.

[Hu, 2001] Hu, Y. H., editor (2001). *Data Transfer and Storage (DTS) Architecture Issues and Exploration in Multimedia Processors*, volume Programmable Digital Signal Processors – Architecture, Programming and Applications, chapter 1, pages 1–39. Marcel Dekker Inc., New York.

[Hüls, 2002] Hüls, T. (2002). Optimizing the Energy Consumption of an MPEG Application (in German). Master's thesis, University of Dortmund, Dortmund, Germany.

[Hwu et al., 2003] Hwu, W.-m. et al. (2003). The IMPACT Research Group. *http://www.crhc.uiuc.edu/Impact/.*

[ICD, 2003] ICD (2003). The ICD Homepage – Embedded Systems Profit Center. *http://www.icd.de/es.*

[Infineon Technologies, 2001] Infineon Technologies (2001). *TriCore 32-Bit single-chip Microcontroller, Architecture Manual V 1.3.1.* Infineon Technologies AG, Munich.

[Intel Corp., 2002] Intel Corp. (2002). *IA-32 Intel Architecture Software Developer's Manual*, volume 3. Intel Corp., Santa Clara.

[Jakubowski, 2002] Jakubowski, J. (2002). Architecture-independent Source Code Optimization using Pattern Matching (in German). Master's thesis, University of Dortmund, Dortmund, Germany.

[Janssen, 2000] Janssen, M. (2000). *Word-level Algebraic Optimisation Techniques for Accelerator Data-Paths and custom Address Generators*. PhD thesis, Katholieke Universiteit Leuven, Leuven.

[Kaibel and Pfetsch, 2003] Kaibel, V. and Pfetsch, M. E. (2003). *Some algorithmic Problems in Polytope Theory*, volume Algebra, Geometry and Software Systems, pages 23–48. Springer, Berlin.

[Kandemir, 2002] Kandemir, M. (2002). A Compiler-based Approach for improving intra-iteration Data Reuse. In *Proceedings of "Design, Automation and Test in Europe" (DATE)*, pages 984–990, Paris.

[Kandemir et al., 2001] Kandemir, M., Ramanujam, J., Irwin, M. J., Vijaykrishnan, N., Kadayif, I., and Parikh, A. (2001). Dynamic Management of Scratch-Pad Memory Space. In *Proceedings of "Design Automation Conference" (DAC)*, pages 690–695, Las Vegas.

[Kandemir et al., 2000] Kandemir, M., Vijaykrishnan, N., Irwin, M. J., and Ye, W. (2000). Influence of Compiler Optimizations on System Power. In *Proceedings of "Design Automation Conference" (DAC)*, pages 304–307, Los Angeles.

[Kernighan and Ritchie, 1988] Kernighan, B. W. and Ritchie, D. M. (1988). *The C Programming Language*. Prentice Hall, Englewood Cliffs, New Jersey.

[Kim et al., 2000] Kim, H. S., Irwin, M. J., Vijaykrishnan, N., and Kandemir, M. (2000). Effect of Compiler Optimizations on Memory Energy. In *Proceedings of "IEEE Workshop on Signal Processing Systems" (SIPS)*, pages 663–672, Lafayette.

[Kuacharoen et al., 2003] Kuacharoen, P., Mooney, V. J., and Madisetti, V. K. (2003). Software Streaming via Block Streaming. In *Proceedings of "Design, Automation and Test in Europe" (DATE)*, pages 912–917, Munich.

[Landwehr, 1999] Landwehr, B. (1999). A Genetic Algorithm based Approach for multi-objective Data-Flow Graph Optimization. In *Proceedings of "Asia South Pacific Design Automation Conference" (ASP-DAC)*, pages 355–358, Hong Kong.

[Lanneer et al., 1994] Lanneer, D., Cornero, M., Goossens, G., and de Man, H. (1994). Data Routing: A Paradigm for efficient Data-Path Synthesis and Code Generation. In *Proceedings of "International Symposium on High-Level Synthesis" (ISHLS)*, pages 17–22, Niagara-on-the-Lake.

[Leupers, 1997] Leupers, R. (1997). *Retargetable Code Generation for Digital Signal Processors*. Kluwer Academic Publishers, Boston.

[Leupers, 2000] Leupers, R. (2000). *Code Optimization Techniques for embedded Processors – Methods, Algorithms and Tools*. Kluwer Academic Publishers, Boston.

[Leupers, 2001] Leupers, R. (2001). LANCE: A C Compiler Platform for embedded Processors. In *Proceedings of "Embedded Systems / Embedded Intelligence"*, Nürnberg, Germany.

[Leupers and David, 1998] Leupers, R. and David, F. (1998). A uniform Optimization Technique for Offset Assignment Problems. In *Proceedings of "International Symposium on System Synthesis" (ISSS)*, pages 3–8.

[Leupers and Marwedel, 2001] Leupers, R. and Marwedel, P. (2001). *Retargetable Compiler Technology for Embedded Systems*. Kluwer Academic Publishers, Boston.

[Leupers et al., 2003] Leupers, R., Wahlen, O., Hohenauer, M., Kogel, T., and Marwedel, P. (2003). An executable Intermediate Representation for retargetable Compilation and high-level Code Optimization. In *Proceedings of "Workshop on Systems, Architectures, Modeling, and Simulation" (SAMOS)*, Samos, Greece.

[Levine, 1996] Levine, D. (1996). Users Guide to the PGAPack parallel Genetic Algorithm Library. Technical Report ANL-95/18, Argonne National Laboratory.

[Liveris et al., 2002] Liveris, N., Zervas, N. D., Soudris, D., and Goutis, C. E. (2002). A Code Transformation-based Methodology for improving I-Cache Performance of DSP Applications. In *Proceedings of "Design, Automation and Test in Europe" (DATE)*, pages 977–983, Paris.

[Loechner, 1999] Loechner, V. (1999). PolyLib: A Library for manipulating parameterized Polyhedra.
http://icps.u-strasbg.fr/polylib/.

[Loechner et al., 2002] Loechner, V., Meister, B., and Clauss, P. (2002). Precise Data Locality Optimization of nested Loops. *The Journal of Supercomputing*, 21:37–76.

[Lorenz et al., 2001] Lorenz, M., Leupers, R., Marwedel, P., Dräger, T., and Fettweis, G. (2001). Low-Energy DSP Code Generation using a Genetic Algorithm. In *Proceedings of "International Conference on Computer Design" (ICCD)*, pages 431–437, Austin.

[Lorenz and Marwedel, 2004] Lorenz, M. and Marwedel, P. (2004). Phase Coupled Code Generation for DSPs Using a Genetic Algorithm. In *Proceedings of "Design, Automation and Test in Europe" (DATE)*, pages 1270–1275, Paris.

[Marwedel, 2003] Marwedel, P. (2003). *Embedded System Design*. Kluwer Academic Publishers, Boston.

[Marwedel and Goossens, 1995] Marwedel, P. and Goossens, G., editors (1995). *Code Generation for embedded Processors*. Kluwer Academic Publishers, Boston.

[Marwedel et al., 2004] Marwedel, P., Wehmeyer, L., Verma, M., Steinke, S., and Helmig, U. (2004). Fast, predictable and low Energy Memory References through Architecture-aware Compilation. In *Proceedings of "Asia South Pacific Design Automation Conference" (ASP-DAC)*, pages 4–11, Yokohama.

[Minkowski, 1896] Minkowski, H. (1896). *Geometry of Numbers (in German)*, volume 1. Teubner, Leipzig, Germany.

[MIPS Technologies Inc., 1996] MIPS Technologies Inc. (1996). *MIPS R10000 Microprocessor User's Manual*. MIPS Technologies Inc., Mountain View.

[Miranda et al., 1998] Miranda, M., Catthoor, F., Janssen, M., et al. (1998). High-level Address Optimisation and Synthesis Techniques for Data-Transfer intensive Applications. *IEEE Trans. on VLSI Systems*, 6(4):677–686.

[Miranda et al., 2001] Miranda, M., Ghez, C., Kulkarni, C., Catthoor, F., and Verkest, D. (2001). Systematic Speed-Power Memory Data-Layout Exploration for Cache controlled embedded Multimedia Applications. In *Proceedings of "International Symposium on System Synthesis" (ISSS)*, pages 107–112, Montreal.

[Motorola Inc., 1996] Motorola Inc. (1996). *DSP56600 16-Bit Digital Signal Processor Family Manual*. Motorola Inc., Document No. DSP56600FM/AD, Austin.

[Motzkin et al., 1953] Motzkin, T. S., Raiffa, H., Thompson, G. L., and Thrall, R. M. (1953). The double Description Method. *Contributions to the Theory of Games – Annals of Mathematics Studies*, 2(28):51–73.

[Muchnick, 1988] Muchnick, S. S. (1988). Optimizing Compilers for SPARC. *SunTechnology*, 1(3).

[Muchnick, 1997] Muchnick, S. S. (1997). *Advanced Compiler Design and Implementation*. Morgan Kaufmann Publishers, San Francisco.

[Niemann, 1998] Niemann, R. (1998). *Hardware/Software Co-Design for Data Flow dominated Embedded Systems*. Kluwer Academic Publishers, Boston.

[Olaniran, 1998] Olaniran, Q. B. (1998). Emulation of the Intermediate Representation in the IMPACT Compiler. Master's thesis, University of Illinois at Urbana-Champaign, Urbana, Illinois.

[Panda et al., 1999] Panda, P. R., Dutt, N., and Nicolau, A. (1999). *Memory Issues in embedded Systems-On-Chip*. Kluwer Academic Publishers, Massachusetts.

[Paulin et al., 1997] Paulin, P., Goossens, G., and Liem, C. (1997). Embedded Software in Real-Time Signal Processing Systems: Application and Architecture Trends. *IEEE Special Issue on HW/SW Codesign*.

[Peymandoust et al., 2002] Peymandoust, A., Simunic, T., and de Micheli, G. (2002). Low Power embedded Software Optimization using symbolic Algebra. In *Proceedings of "Design, Automation and Test in Europe" (DATE)*, pages 1052–1058, Paris.

[Philips Corp., 1997] Philips Corp. (1997). *TriMedia TM1000 Preliminary Data Book*. Philips Electronics North America Corp.

[Pokam et al., 2001] Pokam, G., Bihan, S., Simonnet, J., and Bodin, F. (2001). SWARP: A retargetable Preprocessor for Multimedia Instructions. In *Proceedings of "Workshop on Compilers for Parallel Computers" (CPC)*, Edinburgh.

[PowerEscape Inc., 2004] PowerEscape Inc. (2004). PowerEscape, Inc. Homepage. *http://www.powerescape.com*.

[Püschel et al., 2001] Püschel, M., Singer, B., Veloso, M., and Moura, J. (2001). Fast automatic Generation of DSP Algorithms. In *Proceedings of "International Conference on Computational Science" (ICCS)*, pages 97–106. Lecture Notes of Computer Science 2073, Springer.

[Rimey and Hilfinger, 1988] Rimey, K. and Hilfinger, P. N. (1988). Lazy Data Routing and greedy Scheduling for Application-specific Signal Processors. In *Proceedings of "Annual Workshop on Microprogramming and Microarchitecture"*, pages 111–115, San Diego.

[Ryder et al., 2001] Ryder, B. G., Landi, W. A., Stocks, P. A., et al. (2001). A Schema for interprocedural Modification Side-Effect Analysis with Pointer Aliasing. *ACM Transactions on Programming Languages and Systems*, 23(2):105–186.

[Saad, 2003] Saad, A. (2003). Java-based Functionality and Data Management in the Automobile. *JavaSpektrum*.

[Sinha et al., 2000] Sinha, A., Wang, A., and Chandrakasan, A. P. (2000). Algorithmic Transforms for efficient Energy scalable Computation. In *Proceedings of "International Symposium on Low Power Electronics and Design" (ISLPED)*, pages 31–36, Portofino Coast, Italy.

[Sjödin et al., 1998] Sjödin, J., Fröderberg, B., and Lindgren, T. (1998). Allocation of global Data Objects in On-Chip RAM. In *Proceedings of "Workshop on Compiler and Architectural Support for Embedded Computer Systems" (CASES)*, Washington DC.

[Smith and Holloway, 2002] Smith, M. D. and Holloway, G. (2002). An Introduction to Machine SUIF and its portable Libraries for Analysis and Optimization. *http://www.eecs.harvard.edu/~hube/software/*.

[Stallman, 2002] Stallman, R. M. (2002). *GNU Compiler Collection Internals*. Free Software Foundation, Boston.

[Stan and Burleson, 1995] Stan, M. R. and Burleson, W. P. (1995). Bus-invert Coding for low-power I/O. *IEEE Transactions on VLSI Systems*, 3(1):49–58.

[Steinke, 2003] Steinke, S. (2003). *Analysis of the Potential for Energy Reduction in Embedded Systems using Energy-Optimizing Compiler Techniques (in German)*. PhD thesis, University of Dortmund, Germany.

[Steinke et al., 2001a] Steinke, S., Knauer, M., Wehmeyer, L., and Marwedel, P. (2001a). An accurate and fine Grain Instruction-Level Energy Model supporting Software Optimizations. In *Proceedings of "International Workshop on Power And Timing Modeling, Optimization and Simulation" (PATMOS)*, Yverdon-Les-Bains.

[Steinke et al., 2001b] Steinke, S., Schwarz, R., Wehmeyer, L., and Marwedel, P. (2001b). Low Power Code Generation for a RISC Processor by Register Pipelining. Research Report 754, University of Dortmund.

[Steinke et al., 2002a] Steinke, S., Wehmeyer, L., et al. (2002a). The *encc* Energy aware C Compiler Homepage. *http://ls12-www.cs.uni-dortmund.de/research/encc/*.

[Steinke et al., 2002b] Steinke, S., Wehmeyer, L., Lee, B.-S., and Marwedel, P. (2002b). Assigning Program and Data Objects to Scratchpad for Energy Reduction. In *Proceedings of "Design, Automation and Test in Europe" (DATE)*, pages 409–415, Paris.

[Strobach, 1988] Strobach, P. (1988). A new Technique in Scene adaptive Coding. In *Proceedings of "European Signal Processing Conference" (EUSIPCO)*, pages 1141–1144, Grenoble.

[Sun Microsystems Inc., 2002] Sun Microsystems Inc. (2002). *UltraSPARC III Cu User's Manual*. Sun Microsystems Inc., Palo Alto.

[Texas Instruments Inc., 1999] Texas Instruments Inc. (1999). *TMS320C6000 CPU and Instruction Set Reference Guide*. Texas Instruments Inc., Literature Number SPRU189D.

[Theokharidis, 2000] Theokharidis, M. (2000). Energy Dissipation Measurement of ARM7TDMI Machine Instructions (in German). Master's thesis, University of Dortmund, Dortmund.

[Trimaran, 2002] Trimaran (2002). An Infrastructure for Research in Instruction-Level Parallelism.
http://www.trimaran.org.

[Uh et al., 1999] Uh, G.-R., Wang, Y., Whalley, D., Jinturkar, S., Burns, C., and Cao, V. (1999). Effective Exploitation of a Zero Overhead Loop Buffer. In *Proceedings of "Workshop on Languages, Compilers, and Tools for Embedded Systems" (LCTES)*, pages 10–19, Atlanta.

[Vander Aa et al., 2002] Vander Aa, T., Lauwereins, R., and Deconinck, G. (2002). Optimizing a 3D Image Reconstruction Algorithm: Analyzing the Capabilities of a modern Compiler. In *Proceedings of "IEEE Workshop on Signal Processing Systems" (SIPS)*, pages 246–251, San Diego.

[Verma et al., 2003] Verma, M., Steinke, S., and Marwedel, P. (2003). Data Partitioning for maximal Scratchpad Usage. In *Proceedings of "Asia South Pacific Design Automation Conference" (ASP-DAC)*, pages 77–83, Kitakyushu.

[Wagner and Leupers, 2001] Wagner, J. and Leupers, R. (2001). C Compiler Design for an industrial Network Processor. In *Proceedings of "ACM SIGPLAN Workshop on Languages, Compilers, and Tools for Embedded Systems" (LCTES)*, pages 155–164, Salt Lake City.

[Wagner and Leupers, 2002] Wagner, J. and Leupers, R. (2002). Advanced Code Generation for Network Processors with Bit Packet Addressing. In *Proceedings of "Workshop on Network Processors" (NP1)*, pages 91–115, Cambridge, Massachusetts.

[Wahlström, 1999] Wahlström, J. (1999). *Energy Storage Technology for electric and hybrid Vehicles – Matching Technology to Design Requirements*. KFB Kommunikationsforskningsberedningen, Stockholm, Sweden.

[Weber, 2003] Weber, T. (2003). The Car of Tomorrow. *DaimlerChrysler HighTech Report*, 2.

[Wilde, 1993] Wilde, D. K. (1993). A Library for doing polyhedral Operations. Technical Report 785, IRISA Rennes, France.

[Wilson et al., 1995] Wilson, R., French, R., Wilson, C., et al. (1995). An Overview of the SUIF Compiler System.
http://suif.stanford.edu/suif/suif1.

[Wolf and Lam, 1991] Wolf, M. E. and Lam, M. S. (1991). A Loop Transformation Theory and an Algorithm to maximize Parallelism. *IEEE Transactions on Parallel and Distributed Systems*, 2(4):452–471.

[Wuytack et al., 1996] Wuytack, S., Catthoor, F., Nachtergaele, L., et al. (1996). Power Exploration for Data dominated Video Applications. In *Proceedings of "International Symposium on Low Power Electronics and Design" (ISLPED)*, pages 359–364, Monterey.

[Wuytack et al., 1998] Wuytack, S., Diguet, J.-P., Catthoor, F., et al. (1998). Formalized Methodology for Data Reuse Exploration for Low-Power hierarchical Memory Mappings. *IEEE Transactions on VLSI Systems*, 6(4):529–537.

About the Authors

Heiko Falk is a researcher at the Department of Computer Science of the University of Dortmund (Germany). He obtained his Diploma degree in Computer Science with distinction from the University of Dortmund in 1998, where he specialized in hardware/software codesign and high-level synthesis for FPGAs. Since 1998, he has been working with Prof. Peter Marwedel as a member of the Embedded Systems Group at Dortmund, where he concentrated on research on high-level source code optimizations. Between 1998 and 2004, he performed his research activities in close cooperation with the DESICS division lead by Francky Catthoor at IMEC, Leuven (Belgium). He received his Ph.D. degree in Computer Science from the University of Dortmund (Germany) in 2004.

Peter Marwedel received his Ph.D. in Physics from the University of Kiel (Germany) in 1974. He worked at the Computer Science Department of that university from 1974 until 1989. In 1987, he received the Dr. habil. degree (a degree required for becoming a professor) for his work on high-level synthesis and retargetable code generation based on the hardware description language MIMOLA. Since 1989, he has been a professor at the Computer Science Department of the University of Dortmund (Germany). He served as dean of that department between 1992 and 1995. His current research areas include hardware/software codesign, high-level test generation, high-level synthesis and code generation for embedded processors.

Index

Symbols

#P-complete , 28, 84, 154

A

Abstract syntax tree , 42, 47, 131, 135, 140
Address
 generation unit (AGU) , 11, 167–169
 optimization (ADOPT) , 19, 20, 127–129
Addressing
 code , 126–128, 167, 170, 173, 177, 178, 182
 non-linear $\sim$, 19
Advanced code hoisting , 119–162, 182, 183, 186, 201–206
Affine
 expression , 66, 94, 95, 99, 141, 142
 function , 10, 66, 70
 loop bound , 94–96, 98–100, 141
Algebraic
 cost minimization , 19
 reassociation , 9
Algorithm selection , 4, 17
Allele , 30, 31
Alternative , 64, 65, 90
Application
 multimedia $\sim$, 10, 11, 163, 164
 network $\sim$, 10
 telecommunication $\sim$, 10
Array padding , 62

B

Basic block , 45, 48, 125

C

C transformation tool (CTT) , 21
Cache , 8, 18, 22, 34, 35, 53, 56, 57, 61, 62, 102–107, 128, 154–158, 173–176, 186, 191–195, 201–203, 207–208
Chromosomal representation , 69–73, 83, 96
Chromosome , 30, 31, 70, 72, 73, 83, 84, 96
 length , 33, 72, 73
Circular
 access pattern , 163, 168, 169
 addressing , 167
Code
 generation , 6, 21, 23, 33, 41, 51
 selection , 7, 23, 37
 size , 34–36, 58, 107–108, 111–113, 158–160, 176–178, 182, 184–187, 195–199, 203–205, 208–209
 straight-line $\sim$, 8, 9
Common subexpression , 120–122, 124, 129–142, 159, 160, 186, 187
 elimination , 5, 9, 19, 42, 66, 124–126, 128, 129, 157, 159, 182, 186
Compiler , 3–10, 20, 21, 34, 35, 37, 38, 42–51, 55, 57, 58, 60, 104, 105, 124, 126, 156, 159, 166–169, 172, 174, 176, 181, 184, 186, 187
 known function, 20
Condition , 62, 65–75, 77–82, 88, 90–94, 96, 98–100, 104, 119, 122, 147–149, 152, 183
 affine $\sim$, 67, 68, 73, 81, 142, 143, 147
 loop-variant $\sim$, 65, 66, 68, 142
 monotony of $\sim$, 70–72, 95
 optimization , 62, 69–80, 82, 84, 91, 94–99
 satisfiability , 62, 67–69, 93, 94
Constant
 folding , 6, 9, 169
 propagation , 9, 169

223